Vectors

The Author's Favorite Columns from *FLYING*®

Len Morgan

Photography by
T.L. Morgan

TAB BOOKS
Blue Ridge Summit, PA

FIRST EDITION
FIRST PRINTING

Library of Congress Cataloging-in-Publication Data

Morgan, Len.
Vectors : the author's favorite—columns from *Flying* / by Len Morgan.
p. cm.
Includes index.
ISBN 0-8306-2087-7 ISBN 0-8306-2083-4 (pbk.)
1. Aeronautics. I. Title.
TL559.M68 1991
629.13—dc20 91-26047
CIP

For information about other McGraw-Hill materials, call 1-800-2-MCGRAW in the U.S. In other countries call your nearest McGraw-Hill office.

Acquisitions Editor: Jeff Worsinger
Book Editor: Norval G. Kennedy
Production: Katherine G. Brown
Book Design: Jaclyn J. Boone TPR4
Cover Design: Lance Bush, Hagerstown, MD

For Terry and Kathy

Other books by the author:

The P-51 Mustang
The P-47 Thunderbolt
The Douglas DC-3
The Planes the Aces Flew (with R.P. Shannon)
The AT-6 Harvard
Airliners of the World
Crackup!
The Boeing 727 Scrapbook (with T.L. Morgan)
View From the Cockpit
Reflections of A Pilot

Contents

Introduction

Someone has said flying is the hardest thing to learn and the easiest to do. It is neither. Given sufficient aptitude and motivation, many people can be taught to fly. Learning the art is difficult, but not unreasonably so; performing it is rewarding, but demanding.

Flying was my profession, writing an avocation. Like the itch to fly, the urge to write defies explaining. My paternal grandfather authored 50 books, my mother two successful biographies. A brother wrote several books and edited an excellent magazine for three decades. Forty years ago I bought an old Royal typewriter for $25, hoping a smidgen of their talent had rubbed off. Move over, Ernie Gann.

My articles were about flying, the only subject I knew much about, but they missed the mark. I knew it and so did the editors who sent them back, now and then with an encouraging note. Soon I had enough rejections to paper a DC-3. Enough was enough. I was about to trade the Royal for something useful like a bottle of scotch when a check for $50 came in the mail.

Thereafter it was not downhill in the shade, but the scales began to tilt. My efforts appeared in minor air journals, all of which stood in the broad shade of *FLYING*, "The world's most widely read aviation magazine." I flew for the world's best airline; I wanted my byline in *FLYING*. In time, the editor accepted a short piece, known in the trade as a "filler," then a couple more and eventually some full-length features.

One day (then) editor Dick Collins flew down to Texas. "We'd like you to do a quarterly column," he said.

"What sort of column?" I said.

"Write about anything you like," he said. I wrote two, mailed them to New York, and began another article. A few days later he called, "We're going to run you monthly. We work three months ahead so do another right away." The columns were entitled, "Vectors." That was more than 12 years ago and I haven't missed an issue since.

Freelancing is not for the easily discouraged. A writer might work for weeks without a sale, then get three checks in the same mail. Writing a regular column is a dream come true and all the more so when the editor offers carte blanche treatment. Dick had a suggestion now and then but the monthly topics were of my choosing. Happily, succeeding editors followed the same policy. A more cordial author/editor relationship cannot be imagined.

A pilot does not regard his airplane in terms of size and power and speed. Of course, he must be thoroughly acquainted with its technical side but its operational "numbers" recede in his awareness to be retrieved as needed. Rather, he is absorbed by control feels, cockpit sounds and smells, and what can be seen inside and out from his seat. Every plane has peculiar capabilities and limitations that give it a personality of its own. To describe it by quoting its top speed and wingspan would be like describing its pilot by telling his height and weight.

Most aviation writers are not pilots. Quite naturally their focus is on the machinery of flight. The contributions of these technical writers have been outstanding. But relatively little has been written about the human involvement in operating the machines, about the art of flying, and the life of professional airmen. In my *FLYING* columns that follow, I tried to fill a little of this void.

L. M.
St. Simons Island, Georgia

1

The past and the future

IT WAS A PACKAGE DEAL: sail on the *Queen Elizabeth 2*, spend two weeks in England, return on the *Concorde*. Such a blend of past and future appealed to this lifelong travel aficionado; my last sea crossing had been on a troopship in 1941, and I'd itched to go supersonic ever since Chuck Yeager did it six years later. The fare for the two of us was little more than for a single round-trip on the SST. So why not?

The cab dropped us at Cunard's Manhattan pier on a May afternoon. There she lay, fuel barges alongside. The *QE2* is downright handsome despite her modernistic stack. With a *Queen Mary* funnel and she'd be perfect. The sight of her monstrous black hull and gleaming white superstructure revived memories of long-ago voyages in peace and war. This should be fun.

It was. The lovely lady of Liberty raised her torch against the sunset and waved us through the golden door; it was pitch dark when we dropped the pilot. Then the ship came alive, slowly picked up speed and nosed into the broad Atlantic. It all came back, the stiff breeze along the decks, the subtle vibration, the distant splash of bow wave far below the rail. I stood on the boat-deck, soaking up the feels and smells and sounds of a great vessel under way, hopelessly reinfected with ocean fever.

Although the clocks were advanced an hour each night, the calendar retreated half a century. This was the way it was. Liners were the sole link between the New World and the Old until after the war. We were wined and dined, entertained and pampered for five glorious days, all of it save bar and casino tabs included in the ticket price. Casino? You bet, plus four restaurants, four pools, two banks, seven bars, a sauna and a 525-seat theater, all easily reached by 22 elevators. With 1,850 passengers and 995 crew, the five-star luxury hotel glided eastward at a steady 30 knots. It's more an experience than a trip.

The great North Atlantic names—*Queen Mary, France, United States*—were long since replaced by numbers—*DC-7, 1049, 707, 747*. The creeping pace of ships was frustrating; their gracious way of travel was over, along with the way of life it reflected. Except for the *QE2's* summer schedules, the only way left to cross is by air. Speed is everything, yet it is the only improvement air travel could offer. Compared with a plush liner, the airliner can't come close in creature comforts.

We reached Southampton at dawn on the sixth day. From the boat train I took a last look at the ship that had taken me back in time. Two thoughts occurred: there was a twinge of guilt at having served the industry that ended hers; and, had there been no flying, I'd have been a sailor. It was 1986 again as we raced for London at two miles a minute.

The aviation history buff who visits Britain will think he's died and gone to heaven. For starts, the Imperial War Museum includes 11 aircraft in London plus 102 at nearby Duxford. My wife, whose interest in aeronautics is hardly rabid, was intrigued by a *Concorde* on the Duxford ramp, G-AXDN. The third built, its bare insides are cluttered with the complex gear used in early testing. In 1974, it set an civil transatlantic record, flying to Bangor, Maine, in 2:56. It is surprisingly large, inside and out.

Its sister, G-BOAB, looked surprisingly small parked next to a 747 at British Airways' new Heathrow terminal. In fact, its fuselage is but 28 feet shorter, though you could walk between two of them inside the jumbo. What a thought-provoking sight, that pair poised side by side for the dash to New York. The big Boeing is unmistakably big-load, long-distance while the needle-nosed SST reeks of sheer speed. Each is in a class by itself.

Soon to be retired DC-7B freighter posed next to its jet replacement.

While all flying is special, *Concorde* flying is very special and *Concorde* passengers are very special people indeed as you discover when approaching the counter marked, simply, *"Concorde."* An agent appears out of nowhere to whisk away the bags, handle the check-in and usher you into an exclusive lounge reserved for the very special.

"Already I'm liking this," said my wife. "Stick with me and you'll wear diamonds," I replied, sampling the canapes at the open bar and trying to look nonchalant. The smoked Scottish salmon and sevruga caviar garnished with lemon on melba toast was quite good.

Over there everything is "*Concorde,*" never "the *Concorde.*" "So you're flying *Concorde. . . .*" "*Concorde* flies to Miami." "It's only four hours on *Concorde.*" Well, it's their airplane. I noticed that the New York papers in the lounge were not yesterday's, so they must have arrived that morning on *Concorde*. When in Rome. . . .

Our flight was scheduled to leave London at 1:20 p.m. and arrive New York at 12:20 p.m. That is, the four-hour hop would span five time zones. At 1:00 we heard, "Passengers for *Concorde* may now board." The door is 5′ 6″ high. "Mind your head," said the purser.

You can straighten up inside unless you're a Celtics guard. With four comfortable seats abreast, the subdued gray cabin is quite roomy. My pocketbook-size window in the thick wall provided a scant view, not that there would be much to see once airborne. Our seats were second row from the front, right side.

We pushed back four minutes late and there was an immediate apology from up front for this tardiness. A deadheading pilot on the jump seat handled the PA chores. His blending of technical and general information was a knack few pilots acquire. The highest time British Airways *Concorde* has flown 10,146 hours. Ours, G-BOAB, had flown 9,736:00, logged 2,766 "supersonic cycles" and made 3,559 landings. Our rotate speed would be 199 knots, he said. Interesting.

The British Airways 747 that had occupied the next gate rolled away and we took the runway. "Not to worry," said the PA voice, "We will have been in New York four hours plus when he gets there." Whether or not the sequencing was intentional, it neatly reinforced the smug illusion of superiority that *Concorde* fosters.

Takeoff thrust, burners on, brakes off. It's best to be seated during the next sweep of the second hand. The acceleration is impressive, the view of taxiways flashing past downright unnerving. Let's see, he's got 12,802 feet of runway and he's looking for 199 knots. . . . On we raced, still gathering speed. Any time now, fellow, we're ready back here if you are. . . . Faster, still faster, then the nose rose way up there and we rushed across roads and fields close enough to touch. "Neat, huh?" said my wife as the gear doors closed. I nodded. Pilots are not good passengers.

We held Mach .95 to the Bristol Channel, then with burners on began a slow climb that would continue until we started down for JFK. The flight data displays on the forward bulkheads told the story: Mach .97, .98, .99. Then a hostess blocked my view. When she moved, there it was: Mach 1.02. Supersonic! It soon reached Mach 2.00, there to remain. With autopilots coupled to Mach velocity, we slowly climbed as fuel load lightened. We eventually saw 59,500 feet and 1,310 mph on the displays.

If the QE2 offers a 50-year flashback, does the *Concorde* show the future? I think so. No one who's tried it will ever again be happy with less.

As for the SST's shock wave, remember that 71 percent of the globe is covered with water, and that's what separates us from the booming Far East. If you've ever flown subsonically from way over there to here, you can regard future supersonic travel as inevitable.

We dropped as light as a feather onto 31R at Kennedy after 3:18 of flight, logging 3:44 gate to gate, and the magic was over.

We rode home in a spanking new ordinary jet that seemed hopelessly outdated, and once more it was 1986.

2

For the record

MY FATHER KEPT A DIARY from boyhood, adding a few lines every morning in elegant copybook script. With a fountain pen; no ball points or felt tips for him. I have no idea what he wrote and never will. According to his wish, the 60-odd journals were destroyed after his death.

Diary keeping is more a custom of Europeans than Americans. Dad was English. Some of us Americans who joined the Royal Canadian Air Force acquired the habit from British classmates and I could kick myself for stopping when we transferred to the Air Corps two years later.

It is an easy habit to pick up, and an even easier habit to kick. Daily personal note keeping is a chore. Often it seems a pointless pain in the neck. The temptation is to put it off; days pass and details are forgotten. For example, how did *you* spend this day last week, much less last month? Repeated lapses lead to waning interest and an urge to forget the whole business. Consistency requires self-discipline.

A pilot's logbook is a diary of sorts. Names, dates, ship numbers, fields, times: How strange the bare facts of flight seem years later. Margin notes can add much interest. Who was the captain when we had that near miss at Nashville? When did we quit Love Field for DFW, McConnell for Mid-Continent? Who was the copilot that night over Kirksville when lightning shattered the radome and left us temporarily blinded?

The answers are in these treasured old books along with the names of jump seat riders and celebrities aboard, mention of engine losses, weather problems, and other unusual situations. A four-word note revives details of a hair-raising or hilarious event long since forgotten. Having "liberated" extra copies, I've been able to keep RCAF logs going for 45 years.

The old format accommodates today's single- and multiengine flying as logically as when it was designed in the 1930s. The 8″ x 9″ pages provide ample space for brief comments. One column headed, "Passenger," reflects a day when every trip aloft was considered worth recording. I've faithfully kept it going and its entries have been useful reminders.

Business and government leaders have no choice but to keep meticulous records of meetings, phone calls, interviews, and travel. Such activities are taped and/or noted in writing by staff members. The necessity of maintaining records is obvious: "Mr. Jones, you have testified that on April 25th last year you discussed this memorandum with Admiral Smith. Please recall for the committee the thrust of that discussion."

"Certainly, senator. According to my log, the admiral and I had lunch together. . . ."

The professional pilot who doesn't make notes about a certain unusual occurrence might someday wish he had. Written reports of such matters as engine failure, severe turbulence and passenger injury are required by law and company policy. A less serious problem—a misunderstood clearance, for example—might be resolved on the spot. Then again, it might not. Many the pilot who has been asked for "his side" of an almost forgotten incident.

My boss once asked about a hard landing reported by a fare who was aboard our trip a month earlier. Admittedly, I bottomed out the struts now and then but there was no note about the arrival in question. The copilot recalled nothing unusual or even which of us flew that leg. Help came from the flight recorder, which confirmed a smooth touchdown. No big deal, but I did send the read-out upstairs. I couldn't have the chief wondering if Morgan was losing it.

Problems with passengers had to be carefully handled. We deplaned an unruly woman who appeared to be intoxicated. (We never said "was.") Months later she brought suit for defamation of character, physical abuse, and inconvenience, demanding an incredible amount for damages.

When confronted with our report, including the names of fellow riders, the police, and other witnesses, she reconsidered. Had we been required to testify from memory long after the fact, the outcome might have been different.

On-board recorder tapes can summon up the bare facts of what happened during an incident under investigation and what crew members said. Recordings made on the ground verify radio exchanges, providing they are not "accidentally erased." When questioned, a pilot's account supported by notes made at the time may reveal mitigating factors no tape could capture. For instance, there could be a valid reason for deviating from an assigned route or altitude and subsequent radio discussion might seem to have resolved the matter. But that might not be the end of it; a supervisor might decide to check into it further. Things are apt to go better with the FAA, as with the IRS, when explicit records have been kept.

So much for the logs a pilot must keep and the notes he should keep. How about a diary *per se*, that is, a daily chronicle of personal experiences and impressions? The answer depends on the person. The young man or woman who is enthusiastic about flying should think on it.

At first blush, diary keeping appears somewhat egotistic. After all, is *your* flying particularly interesting? Is it important? Is it worth your best effort? Is is fun? Well, it is to you and that's who you're writing for.

And if it isn't, I'm sorry for you. Why don't you find something that does stir you? A few pros lose interest; it becomes just another job. They put out minimum effort and long to quit. Certainly, the original novelty and excitement wear thin. The flying life entails monotony and frustration to a degree we never expected. Yet most of us counted ourselves among the fortunate until the last landing and resented the age-60-rule. A responsible professional knows when the time has come to take off his wings.

Diary keeping goes back to the Middle Ages. Historical data of immense importance has come from the personal journals of writers who never thought anyone else would read their content. A diarist reveals himself and his times more candidly in his diary than he ever would to his friends. His private observations gain interest over the years and often assume historical value.

Aviation has produced many thousands of books, mostly by nonpilots. Nearly all airmen who have taken pen to paper have tackled the technical aspect; some of their work is excellent. But few who have tried to describe the flying life and its emotional impact have written well. Credit them with trying.

"Pilots are inarticulate," said Beirne Lay, a pilot author who was anything but. Indeed, describing the confusing welter of emotions evoked by flying requires a large talent. Pilots are also lazy. I've known several with the knack who snorted at the idea of writing, and what a shame. They had great yarns to tell.

Interestingly, pilots who have written well were diarists more often than not. Self-consciously, perhaps, but conscientiously, they jotted down impressions, opinions, experiences, and interesting and unusual happenings—the stuff that dredges up memories long after.

The fast pace and rapidly changing nature of the pilot's world are not conducive to quiet periods of reflection. There is no urge to write about what he flies, or where, or the people who share his work. His feelings about the flying life seem of little consequence.

Then it's over and he tries to remember, but his logs list meaningless dates, forgotten places, and the names of faceless strangers.

3

Command decisions

A QUESTION NO ONE SEEMED TO HAVE ASKED during the Challenger hearing: Why were pilots Scobee and Smith excluded from the go-no-go decision to launch? Both were aware that an O-ring failure had almost precipitated a booster explosion on a previous launch yet they were not told of engineers' specific concerns about Challenger's O-rings on that awful January morning. The decision to go despite these doubts was made for them.

The space shuttle is not an aircraft in the usual sense until it reenters the atmosphere, and its pilots are passengers to that point. But they are as interested in the automated rocketry as in the manually controlled glide to touchdown. The boosters are as important as the shuttle's flight controls.

History repeats itself. Attempts to exclude pilots from the decision loop go back to the earliest days of aviation. There has never been a shortage of nonflying experts who think they understood flying better than pilots.

The Post Office established the Air Mail Service in 1918. A year later Congress was assured that the mail was being flown "regardless of weather conditions," an absurd claim considering the aircraft and facilities of that time. Flights were dispatched with no regard for the weather. In one two-week period, there were 15 crashes and two fatalities. Two pilots reported to a field where the visibility was 300 feet. Both refused to fly and were fired on the spot. The airmail pilot of 1918 stood one chance in four of surviving the next six years.

Tension between operators and pilots continued into the first airline years and led to the formation of associations to protect pilot interests, but these scattered groups were powerless to effect real change. My senior captains remembered that time. One recalled a spring afternoon in West Texas:

"There was a squall line to the east, running north and south as far as you could see with dust flying from lightning strikes. The dispatcher phoned telling me to bring it to Ft. Worth." What did he do? "They'd can you for refusing a direct order; you were finished, blackballed. What do you think I did?"

Flaps position shows this 727 was ready for departure.

Some had bitter memories. Said one, "He argued, but the old man said, 'I'm paying you to fly,' so he took off. He's the one they named the airport for." The power of employment life and death rested in the hands of owners hard pressed to break even. Pilots were generally regarded as prima donnas. Anyone who balked simply didn't have what it took.

The weak local groups gave way to the National Air Pilots Association, which led to the present Air Line Pilots Association. Those who think pilots unionized simply to boost wages haven't read history. While the union provided limited protection against "pilot pushing," and federal intervention made their trade somewhat safer, neither solved the question of final authority. Who had the final word? Pilots worked in a world of grays. They still do; they always will. There are no black and white answers to all the ticklish questions of flying safety.

The individual who flies his own airplane is fortunate. If he makes a 180 or diverts or decides not to go at all, he alone is affected. The pilot who flies for hire has other considerations. For example, the airline pilot who decides that New York might fold and heads for Boston knows he is inconveniencing hundreds of passengers aboard his trip and others scheduled to use his aircraft. His diversion turns a marginally profitable operation into a considerable loss.

He must comply at least in spirit with a slew of federal rules that are as open to interpretation as tax laws, reminding himself that the FAA issued the ticket in his wallet, not his management. He is in unwanted competition with other airmen; the tendency is to believe that if others are making it, he can too.

These and other factors are subtle pressures that conspire to influence his judgment—if he allows them to.

The general aviation pilot-for-hire is even harder pressed, being caught in the same plight with little of the assistance enjoyed by the airline fellow and without union support if his judgment is questioned. A company buys an expensive piston twin so its top people can get more done in less time than if they were at the mercy of airline schedules.

The day inevitably comes when the pilot deems it is wiser to wait than go. It might not be easy explaining to a vice president with a major deal at stake that one faulty component has compromised safety, or that it's best to turn back when unwavering airline contrails can be seen overhead. Should he be questioned, he can only put his foot down or stick his neck out. Either way, he might lose. The "damned if you do, damned if you don't" riddle goes with the job.

Federal Aviation Regulation Part 91 reads, *The pilot in command of an aircraft is directly responsible for, and the final authority as to, the operation of that aircraft.* (Another FAR should require a placard bearing those words in every cockpit.)

If a pilot is directly responsible, it follows that he is always the final authority—but it's not that simple. The unrelenting pressure on him is to keep going, due partly to his own reluctance to delay, divert, or cancel.

The typical pro is a responsible individual who very much wants the system to work. He tactfully but firmly resists the efforts of outsiders to usurp his authority. The information and clearances provided him are often worded to sound instructional, which they are not. A pilot worthy of the title wants information, not instructions.

An embarrassment to everyone else on the frequency is the wimp heard asking a dispatcher or controller, "What do you want me to do?" Those to whom he relinquishes absolute control of his flight will point at him if things don't work out. An in-flight problem is solved by grabbing the wheel, not the mike. Such exchanges, heard too often today, were unthinkable 30 years ago.

Old-timers sometimes exercised their "captain's prerogative" to the point of abuse. I flew with more than one hardnose who roundly blistered a tower operator careless enough to issue "taxi instructions." "I already know how to taxi the airplane. What I want is a *clearance* to the runway and then a *clearance* to take off." Such a retort was invariably followed by a lecture to the copilot about a captain's rights and responsibilities. Say what you will about those old vets, they understood the priorities better than some of the weak sisters who inherited their seats.

The picture has altered drastically since DC-3 skippers called the shots. Today's professional cannot barge off VFR, flying any way he pleases. His progress is directed and monitored literally from start-up to shutdown. He must strictly adhere to limits and procedures laid down by experts who understand his aircraft—as a machine, that is—far better than he does. He must operate into congested areas in close cooperation with ground controllers to whom he is but another blip on radar screens, and follow to the letter any clearances he accepts from them.

It is a popular view that today's pilot plays a diminished role; he punches the buttons, sits back and watches the airplane fly itself more precisely than he can. He's just another cog in the wheel. A review of accidents over the recent years of computerized flight suggests that pilots themselves have been caught up in such thinking. Too much reliance on the wonderful machinery and too little questioning of clearances issued by harried controllers who, though well-intentioned, have no real understanding of certain piloting problems—these have led to disaster.

The basics have not been programmed out of the picture. The final authority still sits in the left seat; he remains directly responsible.

One pilot, frustrated by obviously needless vectoring, was overheard:

"With all due respect, sir, am I up here because you're down there, or is it the other way around?"

That's one way to put it.

4

Passengers

ONE JANUARY MORNING IN 1943 an Army C-47 [militarized DC-3] took off from Accra, Gold Coast, and slowly climbed to 9,000 feet. Its destination—Kano, Nigeria, 762 miles to the northeast. In the left seat was a TWA pilot drafted into military service, in the right a green second looie. In back, 28 men sat on metal benches with their backs to the windows. Stacked between them was an impressive load of freight.

The captain had flown DC-3 schedules for three years. His copilot had 162 hours total time, two of those hours in the C-47 learning how to operate the hydraulics and maintain straight and level flight, but not how to takeoff or land. I am sure of the details because I'm looking at his logbook and route manual. I was that copilot.

When I went back, the 28 riders looked up and it hit me. Good Lord, we're responsible for all these people! And, if my skipper becomes incapacitated, we'll all be up the creek minus paddles, an appalling possibility. But all went well that day and until I got enough of the hang of it to pull off a survivable arrival on my own. Thereafter, the load was of no more than casual interest. We worked to get the cockpit there in one piece and gave little thought to who or what was behind the door.

It is a popular misconception that pilots feel a great weight of responsibility for their passengers. The Air Force tested the theory shortly after World War II. As I heard the story from an officer who was there, several T-33 pilots were wired to record pulse rate, respiration, and the other measures of stress, and sent out to buzz the Nevada desert. The assigned route included hairy dashes through canyons.

Then each reflew the route with a rider in back (a volunteer, surely.) Those readings showed increased tension, which suggested pilot concern for passenger safety. That would have been the wrong conclusion. Whether the load was cargo or people made little difference to us overseas or later in airline work.

The hostess uniforms date this shot beneath a DC-7C.

Of course, we were always aware of the faith placed in us, but it was not something to dwell on. Numbers meant little. We felt no greater concern with 440 people in a 747 than we had in a Convair with 44. Let's modify that: The number of passengers was of considerable interest in lean times. Empty seats meant retrenchment, demotions, furloughs. Happiness was a full house.

Certain passengers do come to mind years later for various reasons. There was the elderly German who wanted to see our DC-3 cockpit. He had crewed the Fokker Dr 1 triplane flown by Manfred von Richthofen, the famed "Red Baron," and had a faded snapshot to prove it. He had since been a Texas farmer; I could have kicked myself for not getting his address.

And a certain item of cargo is vividly remembered. We were flying a C-47 freighter back from Florida with a big R-2800 engine in back. It rode on a heavy dolly and was securely tied down—we thought. I was flying. On final approach at Tallahassee, there was an ominous rumbling sound in back and the nose pitched over alarmingly. Together the captain and I crammed on full power and hauled back on both wheels. We hit hard, slightly nose down. As the tail fell there was another rumble as the dolly rolled back. We ballooned, then dropped heavily onto all three wheels and stayed down. The straps preventing forward movement of the engine had broken. Never again did I fly cargo without personally inspecting all tie-downs.

Sen. Lyndon Johnson often flew with us between Texas and Washington, always on a coach ticket, which did not imply frugality. First class fares rode in the rear of our DC-7Cs. Seasoned travelers knew the quietest ride was ahead of the engines, right behind the cockpit. He would take a three-seat row, remove the armrests and spread out his paperwork.

On Nov. 19, 1963, Vice President Johnson rode with us from Dallas to Austin. Three days later we watched Air Force One leave for Washington with President Johnson and the casket of President Kennedy on board. Mrs. Johnson's frequent trips with us continued after she became the First Lady.

Audie Murphy, the much-decorated serviceman of World War II, was a familiar face on our trips. Passenger pilferage was a serious problem for all carriers. Silverware, plates, food trays, towels, blankets, anything not nailed down vanished in astounding quantities. However, we forgave Murphy for stealing one of our favorite hostesses to be his wife.

We carried celebrities—the Harlem Globetrotters, football and baseball teams, Joe Louis, Sugar Ray Leonard, Howard Hughes, Art Linkletter, Spiro Agnew, Chris Evert, Dr. Joyce Brothers, William Holden, and so on. Joan Crawford swept aboard in a full length mink coat, every bit as glamorous off the screen; aging Pat O'Brien rode our military charter back from Vietnam.

"It's partly because of you and James Cagney and Errol Flynn that I'm here today," I said.

"I've heard that before," he said with a grin. "Those were good pictures, weren't they? They don't make 'em like that anymore." I agreed. They've desexed flying, downplayed the fun of it. Most kids today think it's just another job and there is a growing pilot shortage.

One night as we fired up in an Electra, a young lady said she felt ill. An agent escorted her back into the terminal where, an hour later, she had a baby. A snowstorm closed the field after we left, so there would have been no turning back. "You would have had to play midwife," I told my boss when we learned of it.

"The hell you say! As captain, I would have ordered you to take charge," he said, and I believe he would. Even though I'd seen the training film, I was relieved at the way things turned out.

We lost two passengers. One was a lady who seemed alive when carried off the airplane, but was DOA at the hospital. The other was a young executive who was reaching into the overhead locker when his heart stopped. Such incidents required written reports and interviews to the extent you'd have thought we had been to blame.

Visits to the cabin were encouraged in the early days, this pleasantry falling to the captain. As a DC-6 flight engineer I was often sent back to silence door whistling (wet tissues did the trick) or fiddle with the heating switches. Approaching Dallas at dawn on the sleeper run from Chicago, I'd go back and stow the uppers.

Once a purser was helping with a particularly stubborn berth when, just as the latch clicked, there was a muffled yell from within. We dropped it and almost pitched its sleepy occupant into the aisle. He happened to be one Bill Miller, the president of Mid-Continent Airlines with which we later merged. After a suitable cooling off period, I mentioned the incident to him and we both laughed.

Thinking back on the final years, one 747 passenger always comes to mind. The load was Cambodians bound for stateside resettlement. During the tiresome small hours of the longish Guam-to-Los Angeles leg, I went downstairs for a fresh cup. The families slept in tilted seats, babies in arms, kids sprawled all over the place.

A well-scrubbed little boy stood up and watched me intently. As I approached, his father spoke quietly and he scrambled back into his seat. I pointed to the boy and held up four fingers. The father understood and nodded. I beckoned and the boy looked at his father who nodded again. He followed me to the rear galley where the girls arranged a dish of ice cream, probably the first he'd ever tasted. He regarded us with solemn unblinking curiosity as he cleaned the dish.

What a nice little kid he was. Back in my own seat I thought about him—and my bit part in his life. There I was, an immigrant's son, taking him to the land that has been so good to our family. It reminded me of what a wonderful thing we've got going.

Wherever he is, I wonder if he remembers me as I remember him.

5

The Berlin Airlift

ADOLPH HITLER AND HIS MISTRESS committed suicide on April 30, 1945, as the menacing rumble of Russian artillery steadily drew closer to his Berlin bunker. On May 8 at midnight, the unconditional surrender of Germany became official. World War II in Europe was over.

Yet, three years later, the United States and Britain faced the prospect of renewed conflict. Relations with their Russian comrades in arms had deteriorated to an alarming degree. The appalling specter of World War III was real.

Events more easily understood now than then had led to an ominous confrontation. Two months before the surrender, President Roosevelt, Prime Minister Churchill, and Marshall Stalin met at Yalta to discuss postwar plans for the occupation and reconstruction of Germany. While the brutal treatment of German civilians and military personnel at the hands of advancing Soviet troops was no secret, the Allies regarded the Russians as generally cooperative. "I think they are friendly," said Roosevelt. "They aren't trying to take over all of Europe." By then the president was ailing and anxious to avoid discord; he died two months later. Churchill was not so sure of Stalin's true intentions.

The "Big Three" agreed to divide the country into four occupied zones—American, British, French, and Russian. Control of Berlin would likewise be shared; thus, the capital, which was in the Russian sector, was an island in a Soviet sea. A highway, railroad line, and three air corridors 20 miles wide and 10,000 feet high provided access to the city. Because the Russians had agreed to an Allied presence in Berlin, it was assumed that the unrestricted use of these arteries was guaranteed. Significantly, this detail was not confirmed in writing.

Gradually Stalin's true designs became clear. He had no intention of helping rebuild the German economy and no interest in the ultimate reunification of the German people. Having already reduced five European nations to satellites, he now wanted Germany—all of it. First, the Allies had to be forced out of Berlin.

On June 24, 1948, the Russians halted all surface freight and passenger traffic to the western sectors of the city. Furthermore, they served notice they would no longer supply them with food or electric service from their sector.

That the city might be supplied by air was considered "absolutely impossible" by Gen. Lucius Clay, the American commander. However, his British counterpart, General Brian Robertson, thought it might work, at least until the dispute was resolved. The milk for 6,000 Berlin babies came from Russian sector dairies; something had to be done immediately.

On June 26, C-47s with 80 tons of supplies landed at Tempelhof Air Force Base in Berlin. "Operation Vittles" had begun.

While Washington and London debated, Berliners waited, many expecting to be deserted. The world watched. Then President Truman said, "We stay in Berlin, period." Britain's foreign secretary, Ernest Bevin, agreed. "Under no circumstances will we leave Berlin," he said.

It was with bitter irony that the two air forces that had dropped 70,000 tons of bombs on Berlin now found themselves charged with keeping it alive. The odds against success were overwhelming. Simply to survive, the gutted city daily required 4,500 tons of essentials. The RAF could move 750 tons by supplementing its small fleet with commandeered airliners. The Americans had 102 C-47s, each with a two-and-a-half ton capacity. Considering the food stocks on hand, these supplements would be enough for three weeks. Surely the blockage would be lifted by then.

General Curtis LeMay, commander of United States Air Force/Europe, dispatched every flyable C-47 to West Germany. Ground personnel were routed out of bed and told they would be gone for a few days. Mechanics arrived with tool boxes and changes of underwear. The C-47s began shuttling to Berlin on an around-the-clock basis.

One RAF unit mustered was a Sunderland squadron. The next day the first of its large four-engine flying boats touched down on the Havel See, a lake two miles south of Gatow Airfield, with four tons of tinned beef. The 25 British airlines and charter companies contracted sent a variety of types including Hastings, Yorks, Lancastrians, and Tudors.

The initial effort was reassuring to Berliners but sheer pandemonium for those involved. There were countless problems involving maintenance, loading, scheduling, navigation, and communications. There had been no long-range planning for such an operation. Worse, LeMay realized that even the most efficient use of the fleet at his disposal could not deliver the necessary tonnage. He requested four-engine C-54s able to haul 10 tons; President Truman approved.

Captain G.I. Gore called his wife, Kay. "I've been ordered to the east coast. I'll see you in two or three days." In fact, Gore was unaware of his new assignment and would not be home for eight months. That scene was repeated in Alaska, Panama, across the Pacific, and at stateside C-54 bases as pilots and mechanics dropped everything and were airborne within hours. The arrival of this necessary assistance compounded the initial confusion.

If anyone could pull it all together it was Gen. William Tunner who had commanded the wartime "Hump" operation that delivered 650,000 tons across the treacherous Himalayas between India and China. "You might not have liked him," said one pilot, "but you respected him. 'Willie-the-whip' got the job done."

Tunner ordered a second runway at Tempelhof, another at Gatow, and a new field at Tegel in the French sector. Heavy equipment was cut into pieces, flown in aboard C-82s and reassembled. Working 24 hours a day regardless of the weather were 19,000 German civilians. In 89 days, Tegel was operational.

Tunner rewrote standard operating procedures. Henceforth pilots would depart at closely-timed intervals, make one approach and immediately return to base if unable to land. There would be no more hazardous stacking over Berlin. Airplanes would be dispatched three minutes apart in clear weather, five on instruments. Success demanded a high degree of precision flying.

George Hendrick flew B-24s in the 15th Air Force and joined the reserves after the war. He qualified on the C-47 and was recalled to active duty when the airlift began. "I was over there 14 months, almost from the beginning and until it ended," he said. "They sent you out with an instructor to learn the route; after four or five trips to Berlin they turned you loose."

Hendrick was soon transferred to Fassberg to fly C-54s. "There was no C-54 school," he said. "You learned the airplane while flying to Berlin as a copilot, then got a route check as first pilot. A ship would come in, they'd strip it of seats, upholstery, everything, and in 24 hours it was on the way to Berlin with 10 tons of coal in duffle bags, crocus sacks, anything they could find. Coal dust is combustible so the escape hatches were removed to provide ventilation. If you had to go to the john back in the tail, it was like a black dust storm in the cabin. After three trips to Tempelhof you looked like a John L. Lewis miner."

The USAF and RAF flew the north corridor inbound, the USAF and French the southern. All aircraft exited through the central corridor. A string of radio beacons provided en route navigation; instrument arrivals were directed by ground controlled approach (GCA). If a C-54 lost an engine inbound during daylight operation in clear weather, it landed, unloaded and flew back on three.

"It worked like this:" said Hendrick. "We had two assigned altitudes. The airplanes three minutes ahead and behind you were 1,000 feet higher, so you had a six minute separation at your altitude. Under instrument conditions the times were five and 10 minutes. You knew the number of the airplane ahead and heard him report over the beacons. You watched your time and if you crossed a beacon early, you slowed down. The British and French had their own altitudes and flew the same separations.

"We hauled enough gas for two round-trips, plus reserves. If you missed the approach, you immediately returned. When you were overhead your base, they held the next airplane ready to takeoff, fit you back into the pattern and you went back to Berlin and tried again. Tunner had it worked out to a science."

By the third week, 247 aircraft of varying sizes and speeds were flying 2,300 tons a day to the beleaguered capitol; after seven weeks the 4,500-ton daily minimum was achieved. By late August, 100,000 tons had reached Berlin.

On November 26, Russia rejected a United Nations resolution to lift the blockage. Tunner called for more C-54s; 10,000 more pilots and ground personnel were ordered to Germany. The 300,000th ton was unloaded at Tegel in early November.

There were accidents—a C-47 undershot and went into a Berlin apartment, two others collided over their base, a York crashed on takeoff, a C-54 went down near Fassberg, a Navy R-5D in the Taunus Mountains; there would be more before it was over. Only when the three Berlin terminals were completely fogged in did the steady stream of transports cease. "I don't recall what the minimums were supposed to be," said Hendrick. "If the fellow ahead landed, you shot an approach. Many of the GCA controllers were pilots who had reenlisted as master sergeants. It was a bad winter, but they did an outstanding job. I remember that we flew and slept, flew and slept."

In fact, the first pilots on the airlift logged as many as 120 hours a month. One said, "We flew aircraft that should have been grounded. The maximum fuel leakage allowed in the C-54's tech orders was one thing and the dripping we saw was something else. We flew them anyway." Spares were a continuing problem. Enterprising mechanics descended on Gen. Tunner's personal ship while he was at a meeting and "requisitioned" enough parts to ground it for three days. Thereafter, MPs were posted around it."

Winter life in the bombed-out city was sheer misery. There is no count of those who starved or froze to death on nights when the temperature fell below zero. There was scarcely enough food for survival and not nearly enough coal for heating. Still more airplanes and crews were ordered to Germany. By New Year's Day, 1949, 100,000 flights had delivered more than 700,000 tons. The original C-47s were withdrawn as more C-54s arrived.

In spite of the weather, the pace increased, new records falling almost before they could be recorded. The weekly tonnage rose to 41,500 in January, to 44,600 the following month, then to 45,600.

Jim Steele joined the airlift in April following C-54 training in Montana. "It was routine by then," he said. "Those poor sons of guns who were there in the beginning were the guinea pigs. It was trial and error for them, touch and go. The procedures were cut and dried by spring; we had few problems." The "routine" operation recalled by Steele included rolling loaded down a runway without seeing the next pair of boundary lights and landing in near zero-zero conditions.

"It was an exceedingly well designed procedure," said Brimmer Sherman. He flew 144 loads of lifesaving coal and grain to Tempelhof in six months. "There were more than 200 airplanes in the circuit all the time, and another

200 in maintenance. There was GCA at all fields, but the key to the whole thing was good runway threshold lighting."

There were instances of harassment. "A Yak fighter would buzz you or make a head-on pass or the Russians would get on the radio and try to give you a heading that would steer you out of the corridor. In Berlin you stayed with your airplane. Every airlift pilot remembers the mobile canteen that met every crew; that hot chocolate was out of sight. In 15 minutes you were ready to go," said Sherman.

A maximum effort on April 16 saw 12,940 tons arriving in 1,398 flights. Berlin was now receiving more supplies by air than it had by rail before the blockade. On May 10, 9,260 tons set the second high.

It was a humiliating defeat for the Russians. Two days later the blockade was lifted; the surface routes were reopened. The immense effort was continued until late September to build a stockpile of supplies.

The final statistics were staggering: 689 airplanes had flown 124 million miles and delivered 300,000 tons at an estimated cost of $350 million. Sixty-five lives had been lost. A monument to them stands outside the Tempelhof terminal today.

A career USAF pilot, Sherman flew 32 missions in B-29s during the Korean War. "We couldn't cross the Yalu River, we couldn't bomb this bridge or that road for political reasons. We could have had that war over in a matter of months but our hands were tied. It was so aggravating. And we saw the same thing happen in Vietnam. But on the Berlin Airlift we knew what had to be done and we gave it everything we had. No one told us anything but, 'Load it, get it there, then come back for another.' "

The successful supplying of the German capital by air was a turning point in Allied/Russian relations, the importance of which would not be fully understood for years. "The Russians learned we weren't going to back down," said Sherman. "Who knows where they would have gone afterwards if we had? We are there today because of the achievement of the Berlin Airlift."

[Editor's note: Forty years after the blockade was lifted, the Berlin Wall that had separated the city for more than 20 years was razed. Undoubtedly international commerce of the late 20th century, spurred by the speed and convenience of air travel—a worldwide airlift—played a part in Germany's unification.]

6

Big Brother

IN THE DC-3 DAYS a fed rode with us perhaps twice a year. Once, by chance, we had two the same week. My captain was incensed. "We're more closely watched than paroled convicts," he said. As a matter of fact, we were hardly watched at all in those days. We filed VFR, or canceled IFR when clear of clouds, turned off the radios, and leveled off when above the chop. Mild deviations from course were often made to check points of interest or show a passenger his hometown.

As long as we got there in the scheduled time or close to it, no questions were asked and, because the captain kept the log, we always did. When he said we were "on at 10, in at 12," those were the times teletyped to the dispatcher. Captains got two instrument checks and one line check annually, copilots none. Out on the line we were left to our own devices.

These casual ways worked well enough. We tried hard to deliver safe, comfortable, on-time service and, all things considered, the record was good. Despite their nonchalant approach, my captains had a keen sense of responsibility. Supremely individualistic, each had his way of doing things and wise was the copilot who paid attention.

American civil aviation was virtually uncontrolled during the first 24 years of flying. You could build, fly, and repair an airplane to please yourself. That relaxed atmosphere ended with the Air Commerce Act of 1926. The Bureau of Air Commerce was established and granted $84,000 to get the regulatory ball rolling. Traffic rules and licensing requirements for pilots, mechanics, and aircraft were issued in a 48-page booklet. From this acorn a mighty oak would grow. Eventually the bureau's duties were taken over by the Civil Aeronautics Administration, which became the Federal Aviation Administration in 1959.

From the beginning, the causes of accidents were of prime concern. It was estimated that of 178 commercial crashes in 1926, 16 percent were due to "errors in piloting." The rate jumped to 50 percent after a year of federal bookkeeping. Cockpit mistakes have been blamed for most airline accidents in the decades since. Whether this verdict was justified in every instance is arguable.

Pilots soon complained that "pilot error" was becoming the expected, indeed, sought after verdict. Obviously, this conclusion reduced embarrassment and liability for other parties involved—the builders of the aircraft, its engines and components, the airline, the weather, airways and airport specialists and, of course, the government agency which approved the operation to begin with.

Federal promotion of safety had begun with written, oral, and practical tests. Then agents appeared unannounced to ride training flights and scheduled trips. Next came the recording of air-to-ground radio contacts and telephone discussions between pilots and tower and airway controllers. There were occasional complaints that tapes favorable to crews were unreadable on playback while those verifying cockpit mistakes came across loud and clear. Because the feds owned the recorders these suspicions were academic.

Flight recorders noting time, speed, altitude, and vertical acceleration were installed and later replaced by equipment that etched 16 parameters onto brass discs, from rudder position to individual engine thrust. These virtually indestructible devices usually explained how an accident happened, but not why. What was the crew doing and saying in the minutes prior to the accident?

Next came the cockpit voice recorder to preserve radio exchanges and crew conversation. Pilot objection was vehement but the FAA prevailed and a little mike soon appeared in the overhead panel of every airliner. To appease objectors, tape length was limited to the last 30 minutes and the contents could be erased upon arrival providing the aircraft was normally parked and brakes set. Voice and flight data tapes have proved invaluable in accident investigations, yet excessive reliance on them tends to preclude close scrutiny of other evidence.

The hope that tapes would be used solely for investigative purposes was naive, hearings being public affairs attended by reporters hungry for exciting stories. *The Black Box* is a book containing transcripts of cockpit voice tapes from 19 airline accidents, most ending, "Sound of impact." The editor, admitting his own fear of flying, calls them "as dramatic reading as you are likely to find," yet only a professional airman could profit from it. Network television's use of cockpit tapes is an appalling travesty of the "public's right to know." Now, we view a ghastly scene of wreckage while listening to the crew's last words.

On-board recorders might be too damaged to provide useful data; some are never recovered, as when an aircraft goes down at sea. The solution is obvious: radio the data to ground recorders. And, because hearing what was said in an ill-fated cockpit can never be as revealing as seeing what was done, cockpit video taping is the logical next step. This visual record could also be transmitted to the ground. The technology that lets us watch astronauts do weightless somersaults can easily be applied to airlines flying worldwide.

All of which would be brought to pass—if it is—with the ostensible aim of preventing accident recurrences. There is no denying the value of detailed aircraft and crew performance data during accident investigations or that lessons learned from crashes have contributed to increased flying safety. No one wants to turn back the clock, but are we approaching the day when Big Brother will watch as well as listen during every minute of flight? For what purposes other than accident investigations might such tapes be used? The possibilities are chilling.

As a keen observer of the profession for more than 50 years and a privileged member for most of that time, my guess is that in-flight monitoring will be expanded and intensified, possibly by the scenario suggested. The technology is available; the possibilities are all too obvious. Rather than protest its installation, airmen should oppose its misuse.

The likely sensationalized spinoffs of cockpit video tape shown at an open hearing are horrifying. Is the day coming when we not only listen to a crew's last words but witness its futile attempt to avert disaster, all in living color on our den TVs? Surely this would surpass any pretended violence and be the ultimate in vicarious thrills.

It need not come to that. At this writing a senior Washington lawmaker is under investigation for possible breaches of ethics. For reasons not made clear, the hearings are being conducted behind closed doors. He has not been removed from duty in the meantime. Eventually the committee (comprised of fellow lawmakers) will issue its findings. The minutes will not be made public.

That fellow is no more entitled to special treatment than a citizen pilot embroiled in an accident investigation, whether or not he survives to testify. Accident hearings will continue to be open to the public, but the release of cockpit recordings for media exploitation should be forbidden by law. Does the public's "right to know" include the satisfaction of its morbid curiosity? Professional pilots don't have to accept that; they have the clout to confine the use of cockpit tapes to probes for causes, if only they will use their clout.

But that is unlikely. For all the talk about cockpit teamwork, airline pilots remain an individualistic breed slow to unite in a common cause, a trait that repeatedly has harmed them. Characteristically, they complain among themselves and gain little. Meanwhile, general aviation airmen watch, imagining themselves to be unaffected. They should remember that they fly the same airways to the same airports and are part of the same "problems" the FAA is determined to correct by whatever new traffic restrictions and cockpit gadgetry it deems necessary.

7

Salad days

THE CHIEF PILOT THUMBED THROUGH MY ARMY RECORDS, pausing at one entry. "Homestead. You qualified as first pilot?" Air Transport Command's school at Homestead, Florida, trained crews on the C-54, the jumbo of its day. A coveted assignment, Homestead was rough; no airline school would be as demanding. A failed exam washed out a major as quickly as a shavetail. Flight training was as exacting; the hood—similar to a curtain, which blocked the pilot's view outside the cockpit—was put up before brake release and went down at 200 feet on final. It was a memorable six weeks.

"I was a check pilot there," he said. "Be in school 8:30 Monday." The Homestead ordeal had paid off; three years of frustrating job-hunting were over.

Eight of us vets were hired to satisfy a new CAA rule that a DC-6 or Connie crew must henceforth include a flight engineer. The uniform was a double-breasted blue affair with black stripes and brass buttons. It draped like a horse blanket and was so described by old hands. Worse, it was winter weight gabardine, a miserable choice for sweltering southwest routes. Plain bill caps all round; scrambled eggs for captains came later.

The cost of uniform and flight bag would be deducted from our pay. With the prize in hand, it seemed crass to ask what the pay would be. It transpired that we would draw $47.50 a week while in school, less $6.25 for the uniform. Out on the line it would be $62.50, with an increase of $6.25 every six months until $100.00 was reached. That was it, except that DC-4 and DC-6 copilots got $10 and $15 bonuses, respectively. Because it took eight years to advance that far, the bonuses were academic. We'd be on probation the first year and for another year as copilot.

The DC-6 engineer sat on a hard fold-down seat between the pilots. Most skippers accepted the new hand and put him to work. The odd hard-nose resented the implication that he and the copilot couldn't manage between them. With one of them I attained invisibility, being completely ignored.

Excluded from all activity and conversation, I was like a fan at Forest Hills. Eventually the old boy came around and was thereafter a prince to fly with. Some captains were hard to figure.

Our top numbers were in their 40s and would fly until age 65; there was no automatic advancement due to retirements. The management was conservative to the point of paying cash on delivery of new equipment. These factors, coupled with a slow introduction of larger and faster types produced crew positions set in concrete. You were a captain, a copilot, or an engineer and thus you would remain forever and ever, amen—or so it seemed. A senior engineer would get a DC-3 bid in summer, then downgrade when the first snow sent riders back to Pullmans. An accident or two in midwinter meant furloughs. The waltz was three slow steps forward and two quick ones back.

It was back and forth to Chicago, nonstop both ways if you served with one of the "nine old men." There were no "duty rigs," thus no incentive to hurry us home, so we spent long hours sitting in terminals. We flew 85 hard hours a month, often right to the minute. Half a dozen copilots or engineers, each with an hour or two left, would crew an end-of-the-month trip, the junior man working while the rest sat in back reading TIME articles about the big wages airline pilots earned.

Because captains drew extra pay for night flying, the most senior drew the bottom flat-salaried assistants. On the 20th, the next month's schedule patterns were posted. Only after captain awards were published were we allowed to bid. It was a popularity contest. Eventually it dawned on the graybeards that bad luck was not the reason they always drew the greenest assistants.

Thereafter all bids closed together, which solved nothing, for we scanned the captains' choices and adjusted ours; seniority is as valuable in avoiding as in getting. They caught on, put in a new bid box with padlock and thereafter we could only hope for the best.

But not to belabor the point. Most left-seaters were good heads and most months were fun. Make no mistake—they ran the show aloft and on the ground. Because the Airline Pilots Association's bylaws gave a captain one vote and copilots and engineers a half vote each, the skippers' control of policy was absolute. They owned the local council outright and, needless to add, appointed themselves to contract negotiating committees.

It was hardly democracy in action, yet it is unfair to blame it all on the seniors' egoism. The early airliner was flown by a single pilot. When two- and three-engine airplanes were introduced, a second pilot was dictated by law. The new boy was more an apprentice than a "co"pilot, in fact, some lines charged him for the chance to learn at the feet of the masters.

The industry was still young when we were hired (our line was 21) and flying an airplane was still seen as a one-man show. A captain received semiannual checkrides, his assistant none. He bore full responsibility, was on the carpet alone if something went wrong and perhaps deserved a lion's share of the crew budget.

This philosophy had lost its credence by 1950 but continued to pervade;

thus, typical monthly pay for a DC-6 night run was $275 for the engineer, $460 for the copilot, $1,200 for the captain. Fair or not, we joined the game knowing the rules and were prepared to play by them. Senior copilots openly grumbled but we new hires kept quiet. A classmate said, "It's a nice job if you can afford it."

Dallas-based crews also flew DC-4 and -6 schedules south to Peru. An international engineer got an extra $30 a month, well worth going for. We laid over in Panama each way until a multiple crew plan allowed us to fly straight through. Copilots were qualified as relief captains and engineers as copilots, allowing one man to rest on each of the four trip legs while maintaining a full crew up front. My copilot check-out lasted the 15 minutes needed to make three landings.

Until the CAA ruled the scheme illegal, it was interesting duty. At Guayaquil, Ecuador, the fellow off duty would take an unsold upper berth, fasten the belt and try not to think about the forthcoming dash down the gravel strip in a tropical downpour. There was a hill at the end. I remember once being awakened by an awful silence and looking out the little upper berth window—at the Havana terminal. Time to go back to work. Those uppers weren't half bad when you were bushed.

Advancement was painfully slow. The third summer I got time in DC-3s, then it was back to the fold-down seat in October. Gradually the furlough specter faded. It was all too obvious that the safe operation of a DC-6 on all-weather schedules required three crewmen. Copilots and engineers shared the responsibility and should be held partially accountable.

The traditional image of the all-wise mentor with two inexperienced aides began to crumble. (Some of our top copilots had logged 15,000 hours by the time they became captains.) After much infighting during which, to their credit, many captains sided with us, the bylaws were changed to give every member a full vote. Copilots and engineers demanded and received percentages of their captains' pay. Thereafter annual proficiency checks for both ranks were required. The day of the "total crew concept" had not arrived, but it was closer.

The salad days were over, not that it was to be filet mignon from then on. At least we junior fellows could trade in the beat up old Chevy and actually go somewhere on vacation. I'd gone through my wartime savings and borrowed the limit on insurance. Living from payday to payday was depressing, no matter how nice the work. For all our sleepless nights, my small class had been spared furloughing. Now things would be somewhat easier.

We warmed engineer and copilot seats for 16 years before the chance to wear four stripes. Thirty years from the month we were hired, a friend and I sat down in our final groundschool class. "Back when we were DC-6 engineers, did you dream that one day you would fly as captain on the world's largest airliner?" I asked.

"We've paid our dues," he said.

8

Ignition!

PILOTS IN THE 1930s often spoke of engine-airframe combinations. A Curtiss Robin mounted an OX-5, which was a water-cooled V-8, or a Challenger, which was a six-cylinder radial. Popularly they were called "OX-5 Robin" and "Challenger Robin." Also the "OX-5 Travel Air," "Hisso Travel Air," and "J-5 Travel Air," which were similar airframes mounting the Curtiss OX-5, Hispano-Suiza, and Wright J-5, respectively.

Others were "Warner Fairchild," "Ranger Fairchild," and "Kinner Fleet;" the last combination was curious because virtually every Fleet had a Kinner radial.

The custom did not extend to airliners. They didn't say, "Conqueror Condor" and "Cyclone Condor" or "Wright DC-3" and "Pratt & Whitney DC-3."

Pilots were more anxious about powerplant health then compared to now, which must explain it. We expect today's engine to perform as advertised, if properly maintained and serviced. Many the pilot flying now who has never experienced an engine failure. Indeed, the reliability of modern engines fosters complacency, the cause of many tragedies yesterday's pilot would have survived. He expected failures; engines monopolized his attention; forced landings were part of the game.

I've heard old-timers recall those days, raving over one engine, cursing another—their praise or condemnation based on personal experience. For reasons only they can explain, the Curtiss OX-5 is remembered with fondness for all the gray hair it produced. Many clubs have been based on favorite airplanes, but the only organized engine buffs I've heard of belong to the OX-5 Club.

If there isn't a Rolls-Royce Merlin Club there should be. What a magnificent engine that was. I remember the starting drill: mixture rich, fuel pump on, crack the throttle, ignition on, lift the switch. Four or five prop blades slowly swung past then, with an abrupt shudder and sharp rifle crack, it came

Ancient instruments in a still-flying old Waco UPF-7.

alive, spitting flame and smoke from 12 short stacks. It was instantly clear you were strapped to no ordinary flying machine.

A question hotly debated prior to World War II was: which would be the powerplant of the future, the air-cooled radial or the water-cooled in-line? One school held that the easily streamlined in-line would power the fastest types. The radial's frontal area precluded high speeds, they said. Events seemed to support them.

In 1934, an Italian Macchi-Castoldi racing seaplane set a world's speed record of 440 mph. It was powered by an unbelievable Fiat V-24 rated at 3,100 hp. That record fell in 1939 to a Messerschmitt Bf-109 that reached 481 mph with a Daimler-Benz V-12.

The air-cooled engine is lighter and less complex, argued radial proponents, further pointing to the in-line's gaping radiator. They theorized that the radial's superior horsepower/weight ratio would offset its drag. While Europeans leaned toward in-lines, the emphasis here was on radials. The war saw both approaches proved immensely successful.

The Rolls-Royce Merlin powered the Hurricane that first flew in November, 1935, and the Spitfire prototype that flew four months later. When mated to the Mustang airframe, the result was "probably the best all-around single-seat piston-engined fighter to be employed by any of the combatants" in the words of a respected British historian.

Our postwar National Guard Mustangs had Merlins built here by Packard under license. Though remembered for its luxury cars, Packard knew a thing or two about aviation requirements, having built 6,500 Liberty engines in World War I and its own designs later. In 1931, a Bellanca Pacemaker with a Packard nine-cylinder diesel radial flew without refueling for 84 hours, 32 minutes, a record that stood until Voyager's global flight more than half a century later.

Considering my transport aircraft background, the first P-51 takeoff was like riding a slingshot. Adjustment to its sparkling performance came quickly, however, after which there was no sport equal to Mustang flying. The Merlin thrived on hard running, in fact, the manual recommended takeoff power for a minute during each hour of cruise.

Its coolant system was vulnerable in combat. More than one Mustang shot itself down when debris thrown up during strafing entered its underslung radiator. Coolant temperature bore close monitoring. The Merlin was a delight to operate and inspired pilot confidence.

And, how about an R-2800 Club? Just 10 percent of the pilots who have fond recollections of that Pratt & Whitney masterpiece would form a sizable fraternity. The "R Twenty-Eight Hundred" (for its 2,804 cubic inch displacement), an 18-cylinder, twin-row, air-cooled-radial, enjoyed immediate success in the Thunderbolt and Marauder.

That was only the beginning, it also powered the Corsair, Hellcat, Tigercat, Bearcat, Black Widow, Ventura, and Invader. Among the military transports with R-2800s were the Commando, Packet, Mariner, and Chase C-123. In postwar years it powered the Martin 202/404 series, the Convair 240/340/440 series, and all models of the wonderful Douglas DC-6.

The complete list of R-2800-powered aircraft includes 141 production types, 64 prototypes, and five design projects.

The engine's capacity for absorbing combat damage contributed to the Thunderbolt's incredible success at hazardous ground strafing. From D-Day until VE-Day, Thunderbolts claimed 9,000 locomotives, 86,000 railway cars, and 68,000 road vehicles *in Germany alone.*

Our DC-6 R-2800s were easily brought to life—if skillfully managed. You cracked the throttle, squeezed the starter switches, counted six blades, switched ignition to "Both" and depressed the primer switch. When it fired, you stabilized at 800 rpm using primer fuel, then eased up the mixture while easing off on the primer.

Done just so, the big radial settled into idling with sparse smoking. A hair too much throttle and it went off like a French 75; too much prime and the stacks spewed flame. Cooperation between captain and flight engineer was the ticket. What we new hires dreaded was an impatient skipper who pumped the throttle. You know who caught it for the ensuing thunderclap or fire-wreathed cowl.

The Convair's exhaust was routed through "augmentor tubes," the shrouds of which produced hot air for airfoil anti-icing. These acted as mufflers, giving an idling engine a muted growl. An over-primed Convair belched fire at the wing trailing edge.

But I make too much of starting problems. Once you had the feel, firing up was no sweat. The R-2800's performance and reliability were nothing short of amazing. It ingested tropical rain so torrential we heard it above engine roar. Manifold pressures would sag, then those big Pratts would shake like wet bird dogs and go back to work. Many the hairy night we turned on wing lights to watch them slogging away. They were pure marvels. When you were out there above a raging sea, without a prayer of successful ditching, you acquired that curious fondness airmen feel for a stout airframe and the engines that move it.

The radial vs. in-line engine controversy was strangely resolved in that the proponents of each were both right and wrong. The war years would see both in-lines and radials accomplishing the near-impossible. Interestingly, the unofficial world's speed record for piston-engine aircraft is held by a Hawker Sea Fury powered by an air-cooled radial (520 mph in 1966) while the official record (517.06 mph) was claimed by a North American Mustang with a water-cooled Merlin in 1983.

A personal footnote: The timing of my flying life was just about perfect for several reasons, one being that the hours logged were almost evenly divided between piston- and turbine-engined aircraft. Each engine I depended on was, in its day, the finest conceivable. How fortunate was my lot to have known and operated such superb machinery throughout a working lifetime.

9

Adventures in bad eating

WHEN THEY LIST THE PREREQUISITES FOR PROFESSIONAL piloting, good health, certificates, minimum hours, and so on, there's one thing they always omit: a cast-iron stomach. Without one you won't make it through probation.

In the DC-3 days we fared well, food-wise. Compared with what was to come, we ate like kings. Next to the terminal at San Antonio was a small commissary where airline meals were prepared. At dawn they'd fix us scrambled eggs, country ham, biscuits, fresh made coffee. Seventy-five cents. A good start for a 12-hour workday.

And there were plenty of places for a bite en route. Terminals were small; it wasn't half a mile or more to the coffee shop, even at Chicago Midway. Every field had a restaurant, usually open all night. Meals were prepared to order, not handed to you wrapped in plastic, and served on china, not a cardboard plate, and there was a spoon in your coffee instead of a stick. Prompt service and good food was the rule.

We new hires soon learned the ropes. The chili at Ft. Worth was superb, Ft. Smith's whopping hamburger was outstanding, and Wichita offered better homemade pies than your mother baked. There was a nice lady at Burlington, Iowa, who'd have ready the sandwich you ordered when northbound yesterday, along with a sack of corn from her garden.

Joe Gilbert's place at Kansas City was so successful he opened a steak house upstairs. It was one of the best in a city renowned for its beef and you could watch the airplanes outside. His coffee shop featured a different soup every day. Thursday was navy bean, I still remember. On the freighter, we landed at midnight. While the captain did the paperwork, I'd get two one-quart cartons with crackers, more than enough to get us to Chicago on a winter night.

In those days we laid over at downtown hotels where coffee shops were open all night, or eateries were nearby. Across the street from Houston's Rice

Hotel was Kelly's where a dozen or two on the half shell, washed down with a Lone Star, was ample reward for a long day's flying. We stayed in the King Cotton in Memphis where we couldn't find a good restaurant, and the Albany in Denver where we never found a bad one. Corpus Christi was great for fish. Brownsville's Mexican food restaurants were beyond compare. The hot sauce served at the airport made tears run down your face.

Passengers were served box lunches that weren't bad at all. Fresh sandwiches, an apple, carton of milk, and small salad with Thousand Island dressing. It was always Thousand Island. We said the company bought it by the tank car. Coffee was served in paper Dixie cups; no matter how you held the handles, you burned your fingers. Up front we'd toss the boxes out the window, prompting stern warnings about trash found in the air scoops.

On the DC-4s and -6s, somewhat more elaborate fare was served on plastic trays. Put the little pillow on your lap and the tray on it. Meat, mashed potatoes with gravy, and green peas for lunch. Always green peas. We said the company owned 1,000 acres of peas. The same thing with a slice of pie for dinner. Again, not bad. The boss was Catholic, so on Fridays it was fish for lunch and dinner. With peas both times.

Coffee, tea, milk, and soft drinks were served. No booze. We heard that stronger stuff was served on international runs and that cockpit crews of one European carrier had wine with their meals. We stayed with coffee or hot water for bouillon brewed with our own cubes.

Things began going downhill in the 1960s for several reasons. Larger terminals were built and restaurant franchises issued to chains whose prime interest was profit. Slow service, tasteless food, and high tabs were their trademarks. Why the airlines tolerated such treatment of their customers was a mystery, and all the more so because one of the chains was owned by a major carrier. Considering the long walk from the gate and laggard service, we sometimes left without getting anything.

Worse was ahead in the form of snack bars at which hot dogs and stale sandwiches made up the menu. At least they were closer to our gates. Finally came those inventions of the devil, food machines. When you've eaten a day-old meat loaf sandwich warmed in a microwave oven, you'll know what I mean about heavy-duty personal plumbing. The only thing worse than a food machine that works is one that won't. You hungrily eyed the Baby Ruth and cheese crackers while your quarters fell through. Or, no cup fell out to catch the battery acid dribbling from the coffee maker.

Passengers suffered as well. One of the major lines hired a French chef. The ads showed him holding a tray of exotic pastries. Of course, we hired one. His creations were tasty just out of the oven, but were a mess when warmed more than six hours later. The crew consensus was that he should be sent back to Paris. Coach. And force-fed his reheated delights en route.

A scramble to convert airliners into dining cars ensued. Meal services became even more elaborate with linen napkins and coffee poured from a silver pot. This worked well enough on long hauls, but the DC-3 hostess who had an hour to pass out 21 box lunches now served 44 Convair fares in 45 minutes. Sometimes we circled the field while she collected the trays.

One DC-3 schedule saw us leave Galveston for Houston at 6 a.m. Eighteen minutes, gate to gate. We sat in the HOU terminal—listening to the PA and reading in TIME about how airline pilots only worked 85 hours a month—until 1 p.m. when we flew to Corpus. Three hours later we returned to HOU and laid over. We had been on the job 11 hours but only "worked" two. Such conditions were typical throughout the industry.

After considerable debate, managements agreed to "duty rigs," clauses that produced, without increasing operating expenses, the more sensible scheduling we should have had for the asking. We were busier away from base, got home sooner, and layover times were reduced. Eventually we moved from downtown to airport motels. Arriving late usually meant a closed coffee shop and no other food close by.

Most of us brown-bagged it. The Swiss-on-rye with mustard picked up at Kennedy lost much of its appeal by midnight in Omaha, but it beat machine junk food. When a flight engineer said he had a date at the layover station, we asked, "Can she cook?" Never helped.

As meal services were expanded, freezers, ovens, and expensive coffee makers were installed. The culinary excellence of expensive restaurants was often matched, at least, in first class on long-distance trips. The roast prime rib featured on Pan Am's Pacific division was superb. A KLM steward in white coat fixed eggs and Canadian bacon at dawn in a real skillet, the aroma rousing everyone on board.

Sabena's breakfast croissants smothered in Cornish cream and strawberry jam were a delight. The multicourse feast served aboard our own 747s was the envy of the industry. Only once was it equalled during my considerable deadheading and that was by dinner on a British Airways *Concorde*.

The fare on a typical domestic short haul was something else, ranging from fair to absolutely ghastly. If the vice presidents of cabin service had had to face their lines' coach fare every day, they'd have carried lunch pails. Experienced travelers dined before going to the airport and settled for coffee en route. We were stuck with it, or airport snack bars and machine food for days on end and somehow passed our physicals.

Airline food on domestic schedules, where most of it is served, most often proves that new ideas are not always better ideas. In today's frenzied competition, airlines are offering all manner of inducements to fill their seats. I've a suggestion that would cut expenses, please passengers, and see fewer pilots taking medical retirement: Sack the French chef and go back to DC-3 box lunches.

10

Wrong airport

WHEN A 737 LANDED AT FRANKFORT, KENTUCKY, instead of Lexington, its intended destination, the average airline pilot didn't smirk. He winced. He knows how insidiously easy that mistake is. And, once done, there's no undoing it. An airline's operating certificate names the airports into which scheduled flights are authorized. Once the wheels touch down on alien concrete, the trip must be canceled. Any crew foolish enough to continue to the right field (it's been tried) faces a violation as well as an explanation.

What silly mistake could be more mortifying? How to explain to passengers that they must deplane and complete the trip by bus? What to say to the chief pilot and FAA? You would never live it down and always be the poor soul who landed at the wrong airport. There's no chance for a cover-up. Before steps are found to unload puzzled passengers, someone will call the newspaper. The officials upon whose carpet the miscreant is called can often gloss over their blunders, but he screwed up with the world watching. The horrifying specter makes pilots knock on wood.

Landing at the wrong airport is as easy as locking yourself out of the car. At least an embarrassed driver doesn't read about his fishing for the lock with a coat hanger. Our domestic route network is fraught with airport pitfalls. Landing north at Denver in blinding snow, a Convair crew saw the runway, landed, and slid to a stop—at Lowry Air Force Base, five miles south of Stapleton Field.

Racing to beat a thunderstorm to Brownsville one dark evening, a DC-3 landed at nearby Harlingen. The two ex-military fields are remarkably similar in runway layout, so much so that repeats of the goof were occasionally prevented by sharp tower work. The Brownsville controller, hearing a trip reporting on final when there was no one on final, would politely suggest a recheck of position.

Less easily explained was the Louisville-bound crew that braked to a stop on Godman Field at Ft. Knox, 30 miles away. As we heard the story, the skipper, after a 30-day unpaid vacation, repeated the gaffe on his first trip. If true, he must have found himself in large trouble.

Also difficult to explain was a DC-4 that put down at Offutt AFB rather than nearby Omaha's Eppley Field at noon on a clear day. A wag pinned the Jeppesen plate for the SAC base on the bulletin board with a note attached, "Keep off it!" This dubious claim to fame was more than matched by the DC-6 that landed within sight of Eppley but across the Missouri River on a grass strip in Council Bluffs.

More than one crew inbound to Houston Hobby found itself taxiing past fighters at nearby Ellington AFB. Quick advice from Hobby tower averted some, but not all, repeats of that one.

Kansas City Municipal and Fairfax Airport could confuse the unwary. At least passengers deposited at Fairfax could catch cabs to town and arrive as soon as if they had landed across the river at the right terminal.

Miami has attracted airliners bound for Ft. Lauderdale and vice versa. At Seattle it was surprisingly easy to land where your Boeing was built rather than where the timetable said it would go.

I never poked fun at anyone who landed at the wrong airport. I sincerely felt for him and silently thanked him for reminding me to pay closer attention. Such humiliating misadventures shaped me up more than once. Good pilots owe much to the bitter experiences of unfortunate pilots. As someone said, a smart man learns from his own mistakes, a wise man from the mistakes of others.

The penalties for a wrong airport arrival depended on management and the feds. A first offender might get off with a lecture and letter of reprimand. One line's discipline was a simulator check with approaches to the right airport. A fine and/or time off without pay were possible. A month or more in the right seat at copilot's pay could be the penalty. No matter, it was a bad scene all round.

The wrong airport blues has variations on the theme. A certain airline operated from Denver through Colorado Springs, Pueblo, and Amarillo to points east. Word that Pueblo had been turned over to another carrier didn't register with one crew until it parked and noticed with sudden interest that the baggage carts belonged to the new line. Attempting to save the day, the captain called down to the grinning agent, "You got a mechanic here? We've got a rough engine." It worked, the way I heard it.

Some airlines couldn't get through a month without schedule changes. One crew rode out to Newark only to learn its trip now originated at JFK, a 90-minute cab ride away during rush hour. Meanwhile a second crew appeared at JFK to fly a trip—you guessed it—rescheduled out of Newark. No problem, said the dispatcher; if they swap trips, both can depart on time. But, by the time these directions came off the printer, both crews were rushing to their "right" airports.

There were incidents when the wrong airport was a right one even though was the wrong one. For example, a young DC-3 captain completed his line check and launched on his maiden voyage. Every captain has some vivid recollections three-engine airplanes were introduced, a second pilot was dictated by law. The new boy was more an apprentice than a "co"pilot, in fact, some lines charged him for the chance to learn at the feet of the masters.

Upon returning to base, the new skipper reported to his boss, obviously distraught. "Last night I went to Houston—my first trip—and this morning I came back, nonstop both ways."

"Well?"

"I was supposed to land at Ft. Worth on the way back."

The chief, a reasonable man, admitted he'd once done the same thing. He said he'd "take care of it" and that was that. Better no airport than the wrong one.

Two New York crews were dispatched to pick up trips in Europe, one to fly a 747 out of London, the other to collect a 707 in Madrid. Each crew deadheaded to the right airport, only to find the jumbo waiting in Madrid and the 707 at Heathrow. Such supervisory boo-boos never made the news, which hardly seemed fair.

Though it never generated the notoriety of a wrong-airport landing, the "what airport?" landing was an equal blow to pride. Once in a column I confessed that, as a half-fledged Army pilot, I became somewhat confused in flight and, low on gas, landed at a huge air base without the foggiest notion of its location. But can you conceive of two veteran airline pilots not knowing where they had landed? Read on.

A small airplane arrived on a grass field and shut down at the fuel pump. Two elderly gents climbed out and, after much beating around the bush, came out with it: "Where are we?" Amid gales of laughter they allowed they were lost even though they were senior captains—with our airline.

It was all hugely amusing until the gas boy, who happened to be one of our flight engineers, said, "I fly for (us) and I never heard of either one of you." The befuddled old codgers departed in haste, presumably for the right airport. We learned who they were.

While the average pro completes his career without ever landing at a wrong airport, a fair number came close a time or two. I'm one who did. Certainly, there was more chance of it in the days of uncontrolled fields, primitive navaids and no radar. But, strictly speaking, there was no excuse for it then and there is none now. But that's a bit hard-nosed, don't you think?

When the wonderful screen that serves up absolutely correct answers in microseconds suddenly goes blank, the explanation always is "the computer is down" and everyone shrugs and waits for it to come back up. The computer's periodic memory lapses are anticipated and forgiven. No one laughs, no one gets into hot water, and the newspapers couldn't care less. The wrong airport culprit could use a little such indulgence. His mortifying mistake in itself is punishment enough. And he and all the rest of us are better because of it.

11

The Redfern mystery

NEAR OUR GEORGIA ISLAND HOME is Redfern Village, a shopping center. The name didn't register until someone said it is where Redfern Airport was. Of course. It was named for Paul Redfern, an almost forgotten figure in air history. There's no book about him and only brief mention in histories, yet for a few days he was headline news. He still makes the papers now and then, a subject of continuing mystery.

Redfern was born in Rochester, N.Y., and educated in South Carolina. He soloed from a cotton patch in Columbia, S.C., which today is the site of Dreher High School. A plaque there commemorates the fact. He eked out a living at barnstorming and operating an airport in Ohio. Then he was hired at Savannah, Georgia, to spot rum runners inbound from Jamaica.

Lindbergh's solo hop from New York to Paris in May, 1927, electrified the world. Now anything seemed possible; every pilot wanted a piece of the action. Scores of record-breaking attempts were launched. Redfern had enough experience to be taken seriously when he proposed flying nonstop to Rio de Janeiro. Millionaire Howard Coffin, president of the Hudson Motor Car Company and chairman of National Air Transport, agreed to help fund the attempt providing the takeoff was made from a resort island he was developing near Brunswick, Georgia.

Redfern chose a Stinson SM-1 Detroiter for the flight. A large, ungainly six-place monoplane with a remarkable record, it was powered by the same Wright J-5 radial that served Lindbergh so well. The SM-1 was the choice of so many glory-seekers that most built in 1927 were bought for record-breaking ventures. Redfern put down $12,500 and flew to the dirt strip at St. Simons Island, Georgia.

Eddie Stinson himself supervised the fine tuning of NX773. Fuel capacity was increased to 525 gallons, enough to keep *Port of Brunswick* aloft for 52 hours, and extra compasses were installed. Redfern thought the 4,600 miles could be flown in 50 hours. His calculations were optimistic. The full power necessary simply to keep the overloaded airplane aloft during the first stage would greatly increase consumption and long hours would pass before normal cruising speed was reached.

Yet Stinson, eager to see his popular design make still more history, agreed that success was possible. What bothered him was Redfern's refusal to take a copilot. Two days and nights of demanding hand flying surrounded by hypnotic engine drone was too much to expect of any pilot. Redfern was adamant; he would go it alone.

On August 25, 1927, he flew NX773 to the beach at nearby Sea Island, ran up the J-5, and topped the tanks. Virtually everyone from Brunswick was on hand. Paul Varner, chairman of the flight committee, climbed aboard. "We prayed together for a moment," he said, "then I left the cabin." The heavy ship rolled along the hardpacked sand, slowly gathering speed and at 12:46 p.m. it rose. Minutes later it disappeared from view, flying low over the sea, and vanished into history.

The next day word came that *Port of Brunswick* had been spotted near the Bahamas flying at 2,000 feet. Then silence. At Rio the airfield floodlights remained on all night in case he arrived early. Fifty hours passed, then 55 and no further reports. Redfern was down. The few airplanes available flew along his planned route across British Guiana and found nothing.

A radioless freighter reached Jamaica days later. Redfern had circled the ship when it was 165 miles north of Venezuela and dropped a note: Please turn to face the nearest land. The search was renewed, again without success.

An engineer in Cuidad Bolivar, Venezuela, said a green-and-gold airplane had crossed the city leaving a thin trail of black smoke and flown away to the southeast. He copied the registration and large "Brunswick to Rio" lettering on the fuselage, details he could not have known beforehand. Further searches proved fruitless.

Still later, Jimmy Angel said he had seen Redfern's Stinson down in the jungle. Angel was a flamboyant adventurer, a pilot of considerable accomplishment, but somewhat of a braggart. Few took him seriously. (His claim in 1935 to have discovered an unbelievable waterfall was also ignored until surveyors confirmed the drop to be 3,212 feet, the world's highest, and named it Angel Falls.)

That might have ended the story except for reports that a white man was being held prisoner by headhunters who had seen his fall and made him their god. Rumors spread that the man was a cripple due to crash injuries, natives had spoken of such a captive king and so on. A southern state senator even claimed to have found Redfern and talked with him. Thirteen expeditions were organized to solve the mystery and rescue the pilot if he survived. None returned with solid evidence.

As a schoolboy Robert Carlin, the respected aviation historian and artist, became fascinated by the mystery. After completing a World War II tour in B-24s, he ferried his bomber home across the Atlantic and South America. "Flying over British Guiana I went down to 50 feet with Paul Redfern in mind," he said. "That dense jungle stretches away into a purple infinity. There are no huts or trails, just occasional glints of light from streams. It turned something in me. 'He's down there somewhere,' I thought."

Carlin read everything he could find on the subject: old weather reports, Stinson factory records, newspaper accounts. He talked with Redfern's wife, sister, and friends. During five trips to Venezuela he made numerous flights over possible crash sites. His research made him the leading authority on the Redfern mystery. His painting that depicts the probable last minutes of the flight hangs in a private collection.

Carlin believes this is what happened: Redfern's night over the sea was a nightmare. Ships in the area reported intense thunderstorms and water spouts. Miraculously he came through; dawn found him bone weary and hopelessly lost, most likely due to malfunctioning compasses.

He was not badly off course when the Norwegian skipper pointed him toward land. He entered Venezuela at the delta of the broad Orinoco River and mistook it for the Amazon due to inaccurate charts and less than meticulous study of the route. He flew upstream to Cuidad Bolivar, realized his mistake and turned southeast. "At least, this made him the first to fly across the Caribbean," said Carlin.

Valve problems, a common J-5 malady, accounted for the trail of smoke. In fact, Redfern was in more serious trouble than he knew for his engine failed minutes later and he stalled into the trees about 50 miles away and probably was killed in the crash. Even had he survived uninjured, there was not the remotest chance of his walking out of such an unmapped, impenetrable jungle infested with venomous snakes and insects.

Carlin says the "god of the headhunters" stories were pure bunk, "cruel hoaxes that only rekindled the Redfern family's hopes." He is sure Jimmy Angel found the wreckage and pinpointed its position. "I talked with his wife, Marie, and she herself saw it twice. Her story checked out in every respect. I believe I know where the wreckage is within four miles." Will the *Port of Brunswick* ever be found? "Yes, someday," said Carlin. "I only hope they realize what it is."

On Saturday, August 27, 1988, two days after the 61st anniversary of the takeoff, there was a fly-in at McKinnon Airport on St. Simons Island to commemorate the event. Local Experimental Aircraft Association members invited anyone who was interested. "We hoped Robert Carlin could make it," said Bill King, president of the local EAA chapter. "The owner of an SM-2, smaller but similar to the SM-1, came down from Atlanta. There was no air show, except the old Stinson flew along the beach past Sea Island. It was a fitting tribute."

12

It began with bonfires

IMAGINE IT'S A BONE-CHILLING JANUARY EVENING in New York and you are scheduled to fly west across the Alleghenies. The valleys are socked in but an observer down the line says the moon is dimly visible through the overcast. Your destination is reporting broken clouds with occasional light freezing rain. The winds aloft are certain to be on the nose; their strength is anyone's guess. You ponder the weather, route, and equipment, reflecting on past trips good and bad. You fervently hope the navigational aids are working tonight, the bonfires that alone can verify position.

Such was the life of the airmail pilot. Despite unreliable aircraft, he flew with surprisingly regularity when he could see the ground. Flight at night was perilous, flight in cloud impossible. There were no blind flying instruments, radio communications, ground navigation aids, or trustworthy maps.

Navigation was a minor concern in the first days of powered flight. The "aeronaut" flew in ideal weather. If he got lost, he landed in a pasture to get directions. The World War I pilot was sent solo to memorize landmarks, then led on patrol by veterans who knew the way home. Many the disoriented recruit who landed at a friendly field to ask directions. The luckless fellow who confused east with west found himself being toasted in an enemy mess, then dispatched to a POW camp.

Early airmen soon learned that flight in cloud led to loss of control. Providing the machine held together, the subsequent spin or spiral was not always disastrous. Of course, you had to break clear of clouds with sufficient altitude for recovery. Despite the lack of flight instruments, night missions in the first air war were completed with fair success. Although the war saw amazing technical progress, for all practical purposes the airplane remained a clear weather, daylight means of transport. Its sole navigational aid was a compass.

In 1918, the Post Office proposed flying priority mail in surplus warplanes, the most widely used being the DH-4 with a Liberty V-12 engine. At first, the mail was flown by day and entrained at sundown; the time saved was scarcely worthwhile. A government official dismissed the idea as "an impractical fad that has no place in the serious job of postal transportation." Congress was averse to funding a service that ceased after dark and whenever it rained. Success lay in around-the-clock flying, something never previously attempted anywhere.

In a country as vast as ours, it was a bold experiment. There are few areas here today from which no signs of life are visible from 5,000 feet on a clear night. In 1920, the population was a third of what it is now; cities and towns were smaller and less well lit. There was no rural electrification and little road traffic after dark. On a moonless night the mail pilot could fly an anxious hour without a solid clue as to position, his compass virtually useless in turbulence. Bonfires lit by cooperative farmers were the first ground navigation aids.

On route segments flown at night, rotating light beacons were erected at 10 to 15 mile intervals, each flashing an identifying code. Lighted emergency fields were built about 25 miles apart and powerful flood lights installed at terminals. By 1925 the first lighted airway had been extended from coast to coast, a distance of 2,665 miles. It comprised 18 terminals, 89 emergency fields, and 508 beacon lights.

Letters mailed in New York reached San Francisco 36 hours later, half the time required by trains. This included 15 stops for fuel and relief pilots. The eastbound time was five hours fewer due to prevailing winds. By 1933, the federal airway system had grown to 18,000 miles of "light lines" made up of 1,550 beacons and 263 intermediate fields.

The lighted airway solved the problem of night navigation but was only an extension of contact flying. Weather was avoided until the first flight instruments were introduced. Even with them, airmail flying remained hazardous work. By 1927, when the Post Office transferred control to private contractors, 43 fatalities had been recorded.

In that year, the first low-frequency radio range was commissioned. A range station transmitted the Morse code letters "A" (dot dash) and "N" (dash dot) in opposite quadrants. Where the quadrants joined, the two signals meshed into a steady hum, this being a course that could be identified and followed. These spoke-like "legs" or "beams" were aligned along airways and provided fairly direct routes between terminals. It all seemed quite marvelous at the time.

The range network, flight instruments, air-ground radio, and, later, deicing equipment put transport flying on a *most* weather basis, but there were deficiencies. Range legs swung in mountainous areas, near ore deposits and, along with voice communications, were drowned out by snow or thunderstorm static. A steady hum verified being on course but not the distance to the station. A rapid build of signal strength followed by a sudden fade confirmed passage through the "cone of silence" above a station, at which time a pilot computed ground speed since the last station and estimated time over the next.

Expert range flying was an art. Complex orientation procedures had to be memorized and demonstrated on checkrides. After the usual air work, an instructor would say, "Take us home." First, you identified the quadrant, then the leg, and after waiting for a change in volume, the course to the station. Upon reaching the "high cone" you descended outbound, leveled for the procedure turn, then resumed the descent inbound. At the low cone you dropped the gear, descended to minimum altitude (400 feet in most cases), watched the clock, and prayed the runway would be over the nose when the hood [a curtain that covered the windshield] came down.

Hand flying a sluggish old DC-4 through such an exercise while fiddling with power, trim, and range volume control demanded intense concentration. With an engine or two shut down, you were busy. Out on the line, picking signals out of snow static at a mountainous station like Reno was thought-provoking. Most pilots who flew the old low freq ranges are now retired.

The automatic direction finder, like the range, worked best when needed least. At dawn and dusk "twilight effect" made the needle wander aimlessly. In certain conditions it pointed to mountains, electrical storms, and even stations 50 kilocycles removed from the one selected. Most of the time the ADF was a good aid for en route flying and instrument approaches.

Dramatic improvements came in a rush after World War II. The almost static-proof VOR allowed 360 legs [radiating away from the station] to be selected. Course was maintained by watching a needle instead of monitoring the beam's hypnotic drone. VHF air-ground communications were a godsend. During the 1950s, the light lines, emergency fields, and range stations were decommissioned, ending a fascinating chapter of air transport history.

Radar vectoring eliminated tedious shuttling prior to approaches and the ILS reduced landing minimums to 200 feet. With its antenna tilted down, airborne weather radar is useful when approaching a shoreline, or trying to steer clear of one, as when flying down the Russian coast between Alaska and Japan.

Celestial navigation on overwater routes was replaced by Doppler and early loran sets that, despite their shortcomings, did not depend upon clear skies for sextant shots. Then came the inertial navigation system so precise that it notices a change in position during push back from the gate.

What a distance we've come in just 70 years. The modern airliner and business aircraft can be programed to climb, cruise, hold, land, and brake itself to a stop without its pilots touching the controls. They watch, aware of possibilities that cannot be entirely programed out of the picture, ready to intervene when the near-perfect automated mechanisms fail. But they don't fret about bonfires.

13

Landing

NOTHING ABOUT FLYING GETS MORE ATTENTION THAN LANDING. Nothing provokes more analysis, discussion, and training—nor more jokes. No matter what you learn to do in flight, returning to earth is the bottom line.

My logs record more than 21,000 landings, average for anyone with a similar mix of military, airline, and private hours. Because most of them were in two-pilot types, I took the credit (or blame) for about half. Another 2,000 or so were observed as flight engineer and passenger.

Like most pilots, I looked for the "secret"—that elusive knack of smoothly rolling it on. Even under ideal conditions, no one got it right every time, or even most of the time, but some achieved better results than I. I say "most" other pilots because some made little effort at really smooth landings. They flew down to the runway and landed, period. Rarely were their arrivals bone-jarring, and as rarely were they greasers. The majority worked hard at smoothness. I was still looking for the secret at retirement.

When conditions are favorable, it is more than nice to conclude things with a gentle scrub of rubber on concrete. This is easy on the machine, passengers' nerves, and it strokes a pilot's ego. But conditions are not always favorable; a deliberately abrupt return to earth is sometimes the safest way to drop the curtain.

Gusting surface winds or shear reports dictate extra speed directly into the flare, occasionally 20 knots or more than normal. If runway length is marginal, this is not the time to seek smoothness. You "plant" it, raise the spoilers, and get busy with reversing and brakes. A final approach flown at minimum speed on a calm day, as when closely following traffic ahead or to allow a departure to get moving, also reduces the chances of a smooth landing. You flare in the wake turbulence of the fellow ahead and do your best; one time you float and the next you drop like a suitcase full of sand.

Equally difficult is landing out of a tight instrument approach. After long minutes of concentrating on a panel two feet ahead, you break out and have seconds to make the tricky instrument-to-visual transition. Rain or snow on the runway take the edge off ground impact but diminish braking effectiveness.

Few runways are as level as lanes in a bowling alley. Many run uphill, downhill, or both, and the lowest bidder poured the concrete. An expertly executed landing on some runways is followed by a ride that makes passengers think half the wheels are off on the shoulder.

Some airplanes were easier to land than others. The DC-6 was stiff-legged and the Electra much more so, but the DC-4, 707, and 747 would serve up sensuous "paintbrush slapping a board fence" finales now and then if you worked at it; none gave you a dime.

The 727 was in-between, particularly the stretched model. Due to the rear-mounted engines, the mains were way back there; raising the nose to check a sinker drove them onto the runway hard. Easing *forward* on the wheel could save the day, believe it or not.

The heavier, the better. When empty, as on a training ride, the 707-300 handled like a glider. Weight distribution was a factor. We would ask the passengers of a half-full DC-3 to sit in the rear, otherwise, it was difficult to get the tail down for braking. Even in larger equipment, the leeway between center-of-gravity limits could make the same airplane noticeably tail heavy on one trip leg and tail heavy the next.

There were almost as many ways to land a DC-3 as there were DC-3 captains. Most put it on tail low which gave the best shot at a squeaker. One who

DC-3 on short final at Love Field, Dallas.

made consistently good landings preferred a faster approach; he touched down with the tail up where it was in flight. At remote out stations he was known to coordinate power, brakes, and elevators with such skill that he taxied to the gate with the tail high, then eased it to the ground. On departure, he advanced power against the brakes, raised the tail, and taxied out in flying position. Inevitably, these displays came to the attention of the front office, and that was that.

One skipper three-pointed the -3 and got a nice arrival and short rollout almost every time, but there was no margin for error. When the old girl paid off, you had better be no more than 12 inches high. My attempts to emulate him usually were not as successful, yet if I reached for the throttles with the nose way up there and the control wheel against my belt, he'd say, "Don't let your sportin' blood turn to water!"

Then came the tricycles with the small wheel up front and we learned never to let it touch first, or to pull the nose too high at the flare. Dragging the tail skid required an explanation in writing.

Jets gave less room for individual enterprise. You crossed the fence at the right height and speed and wasted no time placing tires on terra firma. One line taught that glide slope sink rate should be maintained to ground contact; in theory the ground cushion would slow descent and produce an acceptable touchdown. But the ground cushion is fickle; it might ease the bump and it might not. Our instructors taught strategy and left the tactics to us. A slight flare, two or three degrees at just the right moment, slicked it on more times than no flare at all.

Landings were fun; some were funny. As a flight engineer I rode through a bizarre arrival. The runway was short and wet and our captain was high and fast. He planted our DC-6 on the numbers but it wasn't finished flying and we ballooned. He forced it on again and again we ricocheted, and off we galloped across the field with our intrepid leader pulling and shoving at all the wrong times.

The copilot let out a cowboy yell, "Whoa, boy, whoa! Easy, big fellow, easy there!"

The tower said, "I show you on at fifteen . . . er, sixteen . . . er, make that seventeen."

I laughed. The old boy corralled it at the last moment and we ground to a stop. He failed to see the humor in it; you know who got chewed out. Years later I reminded him and that time we both laughed.

A score of us were lined up for departure from a long smooth runway on a beautiful clear windless day. Down final came a 707, seemingly with speed and sink rate right on the money. For whatever reason he lost it crossing the fence and arrived like an overload of bricks. The Boeing's struts bottomed out, its wings sagged, and it all but disappeared in a cloud of tire smoke. After a pause, a girl's voice was heard on the tower frequency: "Gee, I wish our captains would let us land!"

If there's one thing you never did it was brag about your landings. Yet a major carrier did, with predictable results. Its magazine and newspaper ads featured weather-beaten old Capt. Pete Williams (or some such name), one of the line's instructors, and he was saying, "We teach our pilots to bring 'em in like great swans." Pictures showed a swan about to touch down and a 707 about to do the same. Of course, this Madison Avenue nonsense put Capt. Pete's students on the spot. Few good landings are made under such pressure.

A line of us awaiting release watched one of their 707s approaching. Sure enough, its edgy driver tried too hard and terminated flight with a crunch that stopped the clocks and must have registered 6.2 on the Richter scale. Someone on the frequency called in, "That you, Pete?" The harder they tried, the worse they got and the scene was many times repeated.

A friend said, "My last landing will be absolutely perfect. If I miss the first time, I'll go around and try again and keep trying until I get it right. What can they do? Fire me?" Happily, his first attempt was a greaser.

I very badly wanted a greaser for that last one before retirement. Conditions were ideal—no traffic, nice long runway, no wind. There could be no excuses. I cut the power, eased back and we touched. "That's a nine point nine out of 10," said our son from the jump seat. I really appreciated that even though we both knew it was an eight.

14

Bells, horns, and lights

NO ONE KNOWS when the first cockpit warning device was installed, or what it was. During the first days of powered flight, an airspeed indicator, compass, and tachometer were typical cockpit instruments. There was little need for warnings or checklists.

World War I saw the frail mounts of daring sportsmen evolve into remarkable weapons. The Sopwith Camel, the most successful fighter, claimed 1,294 victories in two years. Its varnished wood panel included, along with the basic three indicators, a fuel pressure gauge and an oil flow sight gauge. The twin-engine Handley Page 0/400 bomber, able to deliver a 1,650-pound bomb, had a similar bare-bones cockpit, in fact, a single throttle controlled both engines. A knob on top adjusted individual rpms.

A pilot checked fuel, mags, and controls and observed structural and engine limits, at least he was supposed to. More than one out-of-trim 0/400 crashed on takeoff. Whether this led to a mandatory preflight check or was put down to carelessness is not recorded.

Builders concentrated on improving performance, arranging controls and instruments to suit themselves. A pilot made do with what they gave him. This lack of communication between designer and operator would continue for decades and be a contributory cause of many accidents.

The all-metal Ford Tri-Motor airliner was a big step forward in the 1920s. Four pages of its 110-page manual were devoted to "suggestions" for operation, the rest explaining how to drain sumps, grease wheel bearings, and otherwise keep the ship airworthy. You learned to fly by watching experienced pilots and picked up the fine points on the job.

I once flew a 50-year-old Ford 5-AT. Its cockpit made the DC-3's resemble a spaceship's by comparison. Check fuel, mags, trim, and you were set to go. There was no checklist nor warning placards, which is surprising considering the consequences of taking off with trim in the full nose-up position, where it might be left after landing. The pilot guilty of this oversight desperately

Braniff 747 loading at Honolulu for return trip to Texas.

fought to hold the nose down while cranking a handle behind his seat. "You made that mistake just one time," an old-timer told me.

Then came the retractable landing gear that did wonders for speed but was rough on props and belly if you forgot to put the wheels back down. The first pilot to so blunder would not be the last. Installing a horn that blew when power was retarded with gear retracted prevented most recurrences, though there have been many gear-up arrivals in the years since, not all due to failure of retraction systems.

In World War II military tower phraseology was amended to read, "You are cleared to land, check gear down and locked," and you responded that it was. This admonition plus the horn's strident blare made it difficult to explain why your shiny new P-51 blocked the runway with curled blades and a smashed radiator.

Donald Douglas guessed that sooner or later someone would land his DC-3 wheels up. Its retracted main tires protruded just enough to keep the belly clear of concrete. More than one artist with faulty gear shut down engines on final, toggled props to the 8, noon, 4 o'clock position and quietly rolled to a stop. With belly antenna replaced, the old girl was ready to go again.

Green, yellow, and red bands indicating normal, caution, and danger ranges were added to engine and systems gauges. Airspeed indicators were similarly marked. Warning lights called attention to deteriorating oil, fuel, and hydraulic pressures and temperatures rising above safe limits.

Excessive heat in an engine compartment was announced by a red light and earsplitting bell that could be heard by passengers. "Silence the bell!" was the first item on the emergency checklist. A button disarmed the circuit for the ailing engine.

The gear warning horn was a nuisance during training when engine failures were simulated by retarding throttles. The (unauthorized) remedy was to pull the circuit breaker. Now and then a training session terminated with a wheels-up arrival when the instructor forgot to reset it. This led to a circuitry that silenced the blare but was rearmed when the throttle was advanced.

A slew of warning devices have appeared since the first low oil pressure light. The cockpit of a parked airliner today resembles a tilted pinball machine. As cabin doors and baggage bins are closed and engines come to life, the myriad multicolored warning lights go out. At the same time, an array of visual and aural circuits are armed. An intermittent horn sounds if flaps, trim, and spoilers are not configured for takeoff. A clacker rattles if speed is exceeded; the control column shakes, or is shoved forward, if speed is too low. Taped or digitized voices call attention to altitude deviations and excessive sink rates, shouting, "Pull up! Pull up!" if the divergence continues.

These devices draw attention to minor problems before they become serious, and serious problems before they get out of hand, but they are not substitutes for common sense. To rely on them is foolish, for they are as prone to failure as the equipment they monitor.

It has been tragically demonstrated that a takeoff attempted with flaps up has not always triggered the takeoff warning horn. Nor have full power readings on engine gauges always meant that full power was being delivered.

History proves that experimenting with safety devices is no less foolhardy. Until the advent of hydraulically-boosted flight controls, which are locked by trapped fluid when pumps are shut down, various mechanical locks prevented wind damage. The DC-3's surfaces were immobilized with external locks secured with bungee cords. We were urged to "wring 'em out" before takeoff to ensure that locks had been removed, but once every million times a crew found itself airborne with control wheel and rudder pedals locked, a terrifying predicament.

The -3 often could be kept flying with controls locked, so not every blunder ended in disaster. A few imaginative pilots reasoned that pitch could be controlled with the trim tabs. Roll the trim forward to raise the nose and vice versa. Successful, if harrowing, landings were made in this manner.

The crew chief of a C-47 (the military version) at our African base requested that another ship fly alongside while he hacked a hole with the fire axe to get at the elevator lock bungee. This was easier said than done; the Doug had a tough hide. But while we directed by radio, he found the right spot and, after two hours of hard work, reached out and cut the cord. The lock fell free.

A long handle on the DC-4's cockpit floor was raised to lock controls. It was held in place by a tape reeled down from above the windshield. Could the captain and copilot possibly ignore a red-for-danger tape right before their eyes? They could. The tape was a normal aspect of the ground environment, easily overlooked during a hurried departure, and it was almost invisible at night.

Soon after the war, a DC-4 crashed attempting takeoff at LaGuardia Field. It was theorized that the gust lock was engaged. Two of airline history's strangest tales ensued. Federal investigators flying to the crash scene watched in horror as a DC-4 cruising nearby nosed over and dove into the ground, the first and perhaps last time accident experts were eye witnesses to an airline disaster. A possible cause was that the crew, discussing the LaGuardia crash, had experimented with their own lock—and been unable to disengage it.

These tragedies prompted a similar test aboard a DC-6. A check airman riding the jump seat engaged the gust lock without telling either pilot. The ship began climbing. The copilot trimmed without effect so he rolled in more. The check pilot released the handle and the nose abruptly dipped.

The copilot, whose belt was unfastened, was pitched from his seat. His head struck the overhead feather buttons, shutting down both right engines. Control was regained, engines restarted, and an emergency landing accomplished. The checkpilot forfeited his career and a new regulation was issued: Crewmen must keep belts fastened whenever at the controls.

Warning devices and checklists are valuable contributions to safe operation but the thoughtful pilot reminds himself that they are aids, not answers. They point out problems and make suggestions, but only he can supply the solutions.

15

Strange tales

IF ONLY I HAD KEPT NOTES. During his career a pilot sees and hears of many strange things—humorous, bizarre, tragic. Had I made notes of all I came across, I'd have material for a book like no other. Here are a remembered few.

One of the strangest of all strange tales involved a World War I pilot named, appropriately, Strange. After expending his ammunition at a German without effect, he tried to replace the drum on his Lewis gun but found it jammed. He loosened his belt, leaned forward and found himself outside the airplane, which had entered an inverted spin. "I had been cursing because I could not get the drum off but now I prayed fervently that it would stay on forever," he said. He got a foot hooked inside the cockpit, grabbed a strut, somehow regained his seat, and righted the airplane at treetop level. After the war, he met a German pilot who had seen his close call. The details are in Lt. Col. L. A. Strange's, *Recollections of An Airman*.

Air combat has produced countless incredible incidents. After crashing in France, a B-17 tail gunner crawled from his turret to find himself quite alone. His bomber from the rudder forward was nowhere to be seen. His miraculous survival was reported in *The Stars and Stripes*. An RAF Lancaster tail gunner emerged unscathed from a belly landing to discover the rest of his crew had bailed out. There are many documented cases of aircraft with dead crews, or which had been abandoned, landing more or less intact when the fuel ran out. Another Lanc tail gunner, unable to locate his chute when his ship caught fire, elected to jump rather than be burned to death. He fell 18,000 feet into deep snow, survived, and made a living for himself working in a department store.

More than one Hawker Fury pilot assigned to RAF No. 25 Squadron in the 1930s was shaken to see a pilotless "phantom Fury" flying on his wing. The offhand explanation that it was flown "by the spirit of a dead pilot" was all the more unnerving. The illusion was eventually revealed to the newcomer. By dropping his seat and opening a small gun servicing door, a pilot could fly close formation without being seen.

A relative handful of the many pilots trained to fly fighters downed five of the enemy to became aces, yet some who did claimed five in a day, or even during a single flight. One ace-in-a-day was Navy Lt. "Butch" O'Hare, for whom Chicago's airport is named. Perhaps the most remarkable was RCAF Flight Lieutenant Dick Audet. The Canadian tore into 12 German fighters and shot down five on his first combat mission.

Guinness has credited the most carrier landings to a Royal Naval Air Service pilot. I don't recall the number or his name, only that he died in an auto accident. During World War II, land-based fighters were flown from carriers to shore bases. A P-40 pilot inadvertently dropped his belly tank when reaching for the gear handle. With insufficient fuel to make land, he came back aboard without tail hook or barrier. The same feat was accomplished by a Spitfire pilot. A C-130 has operated from a carrier deck as well as a jet airliner, the Fokker F-28.

From our Egyptian base, the RAF flew a Vickers Valentia, a big twin-engine biplane with four-bladed wood props straight out of the first war. The ungainly monster crept along at 80 mph with 22 troops inside and two pilots outdoors. One night several Spitfire types, under the influence to varying degrees, decided to visit a German fighter base—in the Valencia. They lumbered off in the dark, found the blacked-out field, tossed out bombs, and retreated unscathed. Reconnaissance showed that heavy damage had been

Copilot's view of an Electra's Allison turboprop engines at work.

wrought that rather put top brass on the spot. The whole affair was hush-hush but we heard the culprits got off with a lecture.

A Warhawk pilot was apprehended *in flagrante delicto* pursuing a nurse in a Cairo hotel, neither with a stitch on, and charged with conduct unbecoming an officer. At the court-martial, his counsel relied on a line from King's Regs. The regulation: An officer should always be properly attired for the sport in which he is engaged. Case dismissed. Hear! Hear!

In the same vein, the Mile High Club did not originate during World War II as members think, but 150 years earlier. Dashing stalwarts of France's military balloon corps entertained lady friends during tethered moonlight "voyages" after provisioning their craft with "sweetmeats and spirits." It's right there in early histories.

A transient B-24's nosegear collapsed at our African base. The crew was sent along to its unit. When no one came for the bomber, it was put to work as an ice box, taking off at noon and returning at dusk with its cargo of frosty bottles. We could see it way up there, droning back and forth. One day the commanding general arrived unannounced and the cooler was warned to remain aloft. After the parade was over, the general turned to our CO and said, "Tell that damned beer plane to land!"

At dusk on barren Ascension Island in the mid-Atlantic we watched bone-dry P-38s arriving from Brazil. One didn't quite make it and ditched just off-shore. We watched the rescue boat race out. They never found him. Ferrying was a snap compared to combat yet it extracted its toll. At pre-crossing briefings we were given an appalling list of aircraft down in the jungle or sea. We were urged to look for them but I never heard of one being spotted. Most remain right where everything went terribly wrong.

A B-24 in Alaska encountered a Japanese "Betty," a four-engine seaplane and, after a lengthy battle during which the two ships exchanged broadsides like battleships, sent it flaming into the sea. The battered B-24 limped home, too damaged to repair. We were debating the story years later, wondering if it were true, when another pilot overheard. "It happened," he said. "I was flying that B-24."

At Amarillo I saw a closed hangar guarded by MPs. An officer cracked a door. There sat a gleaming B-29. "Forget what you saw," he said. It was hard to keep a new type secret in those days before military reservations covered half of Nevada. Weeks later a B-29 approached our Tennessee base. A hangar was cleared and guards posted. The giant touched down, wobbled, slid to a stop on its belly and there it sat all afternoon. There was no press mention, of course, but half the city drove past for a look. They had better luck with the B-2.

A senior officer regularly buzzed the field in a P-38, a caper that would have seen the rest of us doing the rug dance. We witnessed his comeuppance. He approached from the east, flat out, skimming the trees. From the north came a P-47, low and fast. The timing was perfect. The Jug flashed beneath the P-38 at midfield and climbed into a vertical roll. The fighters were parked side-by-side. Our colonel marched to the P-47, livid. Down climbed a ferry pilot, all sweet-smiling, curvaceous, five-foot-two of her.

Airline flying was once a casual pursuit. One of our captains flew Ford Tri-Motors in the 1930s. A regular fare was a wealthy Texan who asked to be dropped off at his ranch. Why not? This flag stopping continued until the company president was aboard. When everyone piled off for coffee out on the prairie—well, that was that. Our boss hired him the next week.

About that time the agent who loaded bags aboard our Lockheed Vegas at Oklahoma City quit, thinking there might be a better future in broadcasting. Walter Cronkite.

Do flight crews still "check the tires" as we did in DC-3 days? Actually, we marked them with chalk. The line closest to the ground at the next stop won. It was an absolutely tamper-proof lottery, or so I thought until my skipper began winning every time. I caught the old rascal leaning out his side window as we parked and thereafter insisted that tires be marked on the outside.

One of my chores as a new hire DC-6 flight engineer was to help the hostesses make up the berths on overnight runs. The stories I could tell about those sleeper trips! But that's another story.

16

Not in the book

OUR CONVAIR TRIP WAS READY TO BOARD one evening when an agent came up front. He said a couple would be aboard with their son, a pole-vaulter going to a track meet. He wanted to take his pole. It was too long for the underfloor bins, but would fit in the aisle—that is, if I agreed.

I was a junior four-striper eager to please, so the pole was handed in through my side window, carried back, and secured to the cabin floor. That would have been the end of it had not the parents written the company to say what a prince I was. I hoped the boy did better with his vaulting than I did with my explanation to the chief pilot.

Federal Aviation Regulations read that, "The pilot in command of an aircraft is directly responsible for, and is the final authority as to, the operation of that aircraft." That is deceptively broad. Operation of an aircraft involves more than manipulation of its controls. A PIC is also responsible for what happens on board his aircraft, whether he has one passenger or 500. The professional knows that strange things do happen, some humorous, some not at all funny, and he must deal with them all. During a career, an airline pilot might be confronted with childbirth, illness, death, stowaways, or hijackers. More than one has seen them all.

Federal and company rules about carry-on luggage, seat belts, smoking, drinking, and so on are intended to ensure passenger safety, but they cannot cover every situation. Once in a 707 we were 30 minutes reaching the takeoff runway. Then the senior hostess called to say one passenger wouldn't fasten his seat belt. He said he had flown for years, never once fastened it and didn't intend to now.

The flight engineer went back, but to no avail. We could only return to the gate and deplane the man. He would have endangered himself and others had we run into turbulence and I would have been in violation taking him not strapped in. Who'd want him aboard anyway? We broke ground an hour late.

What can you do about 45 passengers on a 44-seat DC-4 at midnight, everyone with a confirmed reservation? Nothing my captain or the agent said, including the offer of a hotel room and breakfast, induced a volunteer to take the next flight. The captain canceled the trip. When everyone had returned to the terminal, we originated as an extra section with a closer head count.

Once in an Electra an engine fire warning light came on when we shut down at Colorado Springs. It was a false alarm but the circuit had to be checked. I advised everyone that a mechanic would shortly arrive from Denver. That did not satisfy a senior senator who was on board. He said he was en route to Washington on a matter of global import and demanded to know exactly when we would get underway again.

As tactfully as possible, I explained that the problem probably was minor but a lengthy delay was possible. He exploded. That was not good enough. The airline was in deep trouble, I was unfit to command its airplane, we'd be hearing from the FAA and so on, and his raucous outburst was vented at the ticket counter, an embarrassment to everyone. I filed the usual report and heard nothing more about it. Sometime later the lawmaker got into hot water over misuse of funds and was censured by his colleagues in Congress.

It was prudent to jot down the details of such incidents and send the boss a copy. A record of the time, place, flight number, witnesses' names, and circumstances was valuable if, as occasionally happened, the legal department received a complaint months after the fact. When asked for advice on coping with ticklish situations, the stock response from higher up was, "Use your best judgment," which was no help at all.

A note from upstairs read, "A passenger had reported a rough landing at AUS on your Fl. 21 of 5/7. Pls adv details." That was two months previously. My log showed 10 arrivals at Austin in May, but not which of us pilots landed on the seventh, nor did I recall any "rough" landings. The engineering department pulled the flight recorder tape, studied the g-forces, and said the touchdown in question was smoother than average. Those black boxes can work both ways. I wanted to write that passenger but the boss said no.

We pilots held passengers in warm regard. An old-timer said, "I work for them. The company charges them and gives part to me. It's the folks in back who put the groceries on my table." There was more to it than profits. Safety, passenger comfort, economical operation, and schedule maintenance, in that order, were our priorities. Providing the best possible service every time was dictated by professional pride. That was the right way to do things. The rare passenger who didn't behave civilly could spoil an otherwise nice trip for everyone on board. He—or she—was a nuisance to other riders and a worry for us.

Booze brought problems. Until 1955, no alcoholic beverages were served on domestic airliners. The Air Line Pilots Association fought hard to keep the rule but with no more success than it had banning trip insurance counters in terminals. Liquor sales receipts are sizable and insurance companies have clout. For all industry's talk about safety, economics is often the bottom line.

Limiting consumption to two drinks didn't improve the rider who had tossed back a few at the terminal lounge or had a flask. An inebriate was not supposed to be boarded but now and then one slipped by. If the hostesses noticed, we had him removed. Otherwise, increasing cabin altitude usually put him to sleep.

But not always. A soused female rider became boisterous, indeed obnoxious, during a multistop run across the Midwest. The situation went downhill from the beginning. The more the hostesses tried to placate the woman, the louder and more profane she became. Finally, I told her she would be invited to get off at the next stop if she didn't cool it. There was no way she would get off before the end of the line, she replied. That did it. When the huge police sergeant beckoned, she quietly left. I filed a report, fully expected to spend a day off rehashing the incident, but heard no more about it.

Airlines successfully avoid most media attention to in-flight disturbances; there are more of them than make the papers. During the first quarter of 1979 one airline reported 51 problems with unruly riders, seven involving physical abuse of stewardesses. Nor do airlines want crew members filing charges against unruly passengers. The employee loyalty that management expects is not always reciprocated.

Neither is the inter-crew relationship always what it should be. Deservedly despised by flight attendants is the captain who avoids involvement in a cabin problem. This wimp says "It's your headache, you work it out, we're too busy to help."

Flying freight is no sure escape from problem passengers. The pilots of a C-47 in China discovered a 10-foot king cobra in the companionway, upreared, hood expanded, inching forward. The copilot slowly rolled a map into a thin tube, lit the end, and thrust it toward the horrifying reptile. It retreated, the cabin door was closed and wired shut.

Once in a C-47 freighter we picked up an aircraft engine. It was rolled aboard and properly secured—or so we thought. It was my leg to fly. On short final at a fuel stop, there was an ominous rumble and the nose pitched over. With both of us holding our wheels full back, the ship hit hard on the main gear. There was another rumble and the tail slammed down. The 2,400-pound radial had broken its rear tiedowns. At a retired pilots bash 30 years later, I asked that captain if he remembered. He recalled specifics I had forgotten.

You could go for months without trouble. There was not "one in every crowd," at least, not in the crowds that rode our trips. For every passenger who worried us, there were many thousands who accepted the best we could offer without complaint. We were grateful to them for their trust and forbearance when factors beyond our control made it impossible to give the service we advertised. Those who caused problems were a minuscule percentage, which is why they are so easily remembered.

17

The air show question

THE AIR SHOW TRAGEDY at Ramstein air base in West Germany in 1988 raised the question: Why was the Freece Tricolor team allowed to perform a maneuver involving a high-speed pass over the crowd? The disastrous possibilities were all too apparent—and what happened at Ramstein had happened before. When John Derry's D.H.110 broke up at Farnborough in 1952, an engine fell into the crowd killing 28 and injuring 63.

And why did an Airbus with 133 joyriders make a low, slow run at a French demonstration? Miraculously, there were only three fatalities when the A320 stalled into a forest and burned. As results of these avoidable tragedies, West Germany and Denmark banned military shows and President Mitterrand halted airliner demonstrations in France.

But the questioning began before either accident and goes deeper. There is a growing public sentiment that air shows as they are presently staged are mistaken. Critics contend that whatever purposes they serve are far outweighed by their costs in lives and equipment. Obviously, exhibition flying involves risk. The occasional crash is hardly noticed except where it happens, and is soon forgotten except by those who are there. It is the cumulative cost that disturbs many people, insiders included.

The Blue Angels and Thunderbirds have suffered 41 fatalities since 1955. A study by the *Los Angeles Times* revealed that the U.S. services together lost more than 100 lives and a billion dollars worth of aircraft at air shows and other demonstrations during the same period. While losses among nonmilitary airmen have not been counted, research into crashes involving restored military aircraft in private hands is underway. Whatever those losses, they are sure to fuel antishow feeling.

The air show is a tradition dating back two centuries. Thousands gathered near Paris in 1783 to watch, awe-struck, as Pilatre de Rozier became the first human to ascend in a manmade device. Thereafter "aeronauts" drew huge crowds to see their balloons silently rise toward the heavens. The airplane first

Restored North American T-6 performs for Reno Air Races crowd.

earned money simply by struggling into the air. Incapable of doing much else, it began as an entertainer.

When the novelty of flying wore off, competitions for speed, distance, and height were staged. Then, inevitably, came "stunting." Many the spectator who got more than his ticket money's worth in thrills when a daredevil's frail craft collapsed in midair. Among the hapless pioneers was young Charles Rolls whose elevator failed. Together with inventor F. H. Royce, he had just founded a company to build cars.

Although the airplane after World War I was vastly improved, it evoked little interest among military leaders, nor could serious civil use be found for it. However, it could give laymen a taste of flight and was robust enough for aerobatics so it resumed its carnival role. The barnstormer hopped passengers wherever he found a crowd and claimed starvation worried him more than crashing. Bands of these aerial gypsies roamed the land as "flying circuses," giving rides and staging shows. Eventually they flew the first airliners. (The stories they told their copilots!)

Stunting was a sure crowd pleaser but could not alone sustain ticket sales. The aerobatic pilot ran through his repertoire in 20 minutes, after which audience interest waned. Daring variations of standard maneuvers did little to hold attention as few laymen appreciated the risks involved. One specialist flew a complete circle at 200 feet while snap rolling, pausing inverted after each roll, an incredibly difficult feat. Not everyone watched but the spin with which he rang down the curtain drew wild cheers. Laymen always loved a spin, the easiest of maneuvers.

Wing walking, plane-to-plane transfers, pylon races, and even deliberate crashes were added to satisfy the public's lust for excitement. Accidents were inevitable and occasionally spectators died along with performers. During the 1930s itinerant carnivals gave way to large events at which military and civil airmen competed for prize money or flew simply to entertain.

The 1928 National Air Races at Los Angeles drew 100,000 people and featured formation aerobatics by the Navy's "Sea Hawks" flying three F2B-1s, their wingtips tied together with short cords. The Army's "Three Musketeers" flew Boeing PW-9Ds with equal precision. The tiny air services (the Army had but 16 PW-9Ds) thereafter seized every chance to attract taxpayer attention.

After World War II a surplus Mustang or Corsair could be had for as little as $1,500. These high performance mounts lent new excitement to the famous Bendix and Thompson Trophy races and are still flown in competition. The Confederate Air Force and similar groups routinely fly restored warbirds at air shows and there are few among the airminded who have not seen the spectacular Thunderbirds and Blue Angels. The air show remains a favorite spectator sport, as thrilling today as when the first hot air balloons went up.

But does it serve any worthwhile purpose other than to titillate thrill seekers? It is certainly true that new ideas developed during racing in the 1930s influenced warplane designs in the 1940s. The Spitfire directly evolved from designer R.J. Mitchell's inspired Schneider Trophy winners, to cite the classic example. But what can today's designers learn from watching antique fighters in closed-course races?

Builders say the showcasing of their products at Farnborough and Paris is essential to sales. Prospective buyers want a close look at what's available, but are we to believe that multimillion dollar contracts for fighters hinge on 15 minutes of aerobatics?

The flying displays of vintage warplanes and civil antiques, billed as "history brought to life," seems worthwhile at first blush although the repeated losses of historic aircraft here and abroad is disheartening to all who love old planes. An appalling number of rare types, a few of them one-of-a-kind, have been destroyed in air show flying.

The flight demonstrations of the Blue Angels and Thunderbirds have no doubt attracted many youngsters to military service even though the announcers' broad hint that every cadet is taught to fly like the elite sextet overhead does lay it on a bit thick. In fact each team comprises the sharpest of the sharp, chosen after tough competition. Whether or not the recruitment benefits will indefinitely justify the costs in the public view remains to be seen. A Ramstein-like mishap at an American air show would bring on a congressional investigation.

But a repeat of that tragedy here is extremely unlikely considering stringent FAA crowd supervision and flight regulations. A federal observer can and will stop a show if the rules are violated. American spectators in designated viewing areas are far more at risk driving home than while watching an air show. People outside field boundaries can pose problems, however. Aware that motorists park on roads adjacent to the field, the FAA canceled a Blue Angels performance at Dobbins AFB last spring.

Not every air show critic is a bluenose who couldn't care less about flying. Many insiders who care a great deal are uneasy. You don't have to be a veteran pilot to appreciate that much air show flying leaves little margin for error. Sensible modifications to flying routines would increase pilot safety without watering down the entertainment value. And, no matter what other benefits are claimed, entertainment is the prime purpose of most air show flying today.

No one enjoys an air show more than I. The expert handling of aircraft is a visual feast for most of us who are average airmen. But I wince at needless risk taking and, while watching some really wild caper overhead, often think, "Careful, friend!" My enthusiasm cooled somewhat after watching a young pilot buy the farm—literally. The terrible hole he drilled was right beside a barn. It was a sight to sicken, but mostly it was sad.

We are more than 85 years removed from the Kitty Hawk success. Aviation has little left to prove. But it still has some growing up to do.

18

The changing scene

THOSE OF US WHO GOT AIRLINE JOBS after World War II sometimes think we saw the best years, and that the profession has lost its appeal since we retired. Many young pilots are likewise convinced that we enjoyed more satisfying careers than theirs. There is some truth in this, but not as much as any of us thinks. The life has changed, but it is different rather than better or worse.

We retirees forget the changes we saw, most for the better. We began in DC-3s without pressurization, VHF and VOR radios, satellite weather plots, radar, reversing, antiskid brakes, grooved runways, and other developments now taken for granted. We had no duty rigs or realistic retirement, hospitalization, or insurance programs, nor was there automatic advancement due to retirements; our seniors had years left to fly. Seasonal furloughs were routine. Copilots served an unstable, low paid apprenticeship lasting 10 to 15 years.

We retired on jumbos that almost flew themselves. The job was safer, much better paid, and promised a comfortable retirement. We thought we had it made. Then deregulation and merger/takeover mania brought it all crashing down for some, while creating better futures for others. Some pilots lost, others gained; it depended on the name painted on your airliner.

The convulsed 1980s were not unique in airline history. Between 1926 and 1931 six small lines merged as United Air Lines, eight as TWA, and 14 combined as American Airlines. Squabbles over seniority matched anything seen today. Then as now, pilots of failed lines had to start all over again.

There was further upheaval after World War II. Nonscheduled airlines were organized by veterans with more zeal than business acumen. With surplus military transports available at a dime on the dollar, profits seemed assured. The resulting bankruptcies, buy outs and mergers echoed the frenzy of the 1920s and have a familiar ring today.

Meanwhile, the "grandfather lines" assessed the future; few guessed right. Too many pilots were hired, or too few. With tens of thousands of vet-

The Convair 440, a dependable workhorse for years.

erans agreeable to any wage for a seniority number, the hiring picture was utter pandemonium. I was accepted by Eastern on Friday; a telegram came Monday: Pilots were being laid off and my class was canceled. A friend wore the uniforms of four lines before landing a spot with ours.

History is repeating itself. The industry is once more emerging from turmoil. Experts predict American, United, and Delta will be among the four or five megacarriers dominating the international scene by the year 2000, leaving others to serve niche markets. Their guess is as good as any. While a recession would renew the shakeout, the industry's new look is discernible. A pilot with airline ambitions who keeps a close eye on the situation will know which lines to head for and which to avoid.

Is it still an attractive career? Is it worth the work and expense of qualifying? Four points should be considered:

First: Safety. Airline flying was not unduly hazardous when we began—but there were more funerals than we like to remember. Better equipment and improved pilot skills have reduced the risk. If you don't crash on the way to work (two of our people did), the odds are that you'll fly until age 60 and get a nice plaque to hang in the den.

Secondly: Pay. No matter how badly you want an airline job, you will be unhappy earning less than "industry level" pay. Some pilots today do earn less, much less. A senior flight attendant on one line can match the pay of a captain on another. Deregulation made it possible to buy an airline by pledging its assets against loans, tear up labor agreements, and pay whatever the new owners wanted. If workers balk, hire outsiders.

Had pilots stuck together, such schemes would have failed. But, characteristically, they sat back and watched as failures and mergers put thousands of them out of work. Owners took advantage of their weak response to institute multitiered pay scales that produce substantially different pay for pilots doing the same work.

These wild disparities cannot last. There is no longer a pilot surplus, in fact, a critical shortage looms. The law of supply and demand will see the best qualified applicants hired by lines that pay to get the best. The quick-buck operators will have to raise wages to attract what's left. You get what you pay for in pilots as in everything else. To some degree the pay mess will be self-correcting.

The best airlines have good insurance, health, and retirement plans. Loss-of-license coverage takes some of the sting out of a busted physical. Here again, the youngster who asks questions will know which lines are the "best."

Third: The work. Airline flying is a highly-disciplined art involving intensive training and frequent proficiency checks. Every minute of flight is regulated and monitored from engine start to shut down. Deviations from company and federal procedures are not tolerated. To get along, you go along with procedures exactly as published. You fly by the book—period.

It was different in our salad days. We saw little traffic. In fair weather we filed VFR, turned radios off, and enjoyed the ride, now and then making mild excursions to see points of interest. Such flying is but a pleasant memory in airline circles. If you relish flying for its own sake, you'll be able to afford your own airplane. We never could.

Fourth: The life. Pilots today work longer for less pay. The job always demanded more time than outsiders imagined. Writers with little inkling of our trade said we "only worked 85 hours a month." In fact, preflight preparation, debriefing, schools, and checks brought our total work time to the level of ground workers and required more hours away from home. A concentrated duty period was followed by a string of days off. That's still true.

Seniority still decides what, when, and where you fly, your domicile and pay. As seniority accrues, so do options until you can pick your base, equipment, and trips. Few other pursuits offer such choices. And it's still true that you will be a guilty soul trying to prove your innocence if you become involved in an "incident," no matter what you heard in school.

The pilot's image has suffered in recent years. In our day, airline flying was a respected profession, though not all was sweetness and light. There was mild friction with some ground people who grumbled that we were overpaid and underworked. Bosses fumed at our requests for new equipment but we knew they were trying to keep the company alive. Considering the pressures we all felt, relationships were remarkably cordial.

Our cockpit atmosphere of mutual respect and cooperation was unappreciated at the time. In the aftermath of deregulation, militant unionists are paired with strike breakers. Other factors have induced bitterness between individuals who are supposed to work together in the interest of safety. The goodwill we took for granted is sadly lacking in some quarters. This deplorable turn of events is one more reason to thoroughly investigate prospective employers.

The trend today is to denigrate pilots. Builders and operators claim that upcoming computerized aircraft will fly more precisely and be able to land in bad weather untouched by human hands. Traditional problems with components will be solved automatically, they say; navigation and approaches will be accomplished by pushing buttons. There will be less need for skill and good judgment, according to this thinking, therefore lower pilot wages are in order. (Does it follow that we'll see a future accident blamed on "computer error"? Don't bet on it.)

What outsiders think is inconsequential. The only image that matters is a pilot's image of himself, his skill, judgment and sense of responsibility. The pros and cons considered, airline flying still promises a rewarding career for the mature individual prepared to face the new challenges that lie ahead. I'd go after an airline job as eagerly today as I did 45 years ago.

19

The missing link

"GUESS WHAT? I've just found *Air Force One*!"

It was longtime friend Richard Broome calling from his Colorado studio. One of the rewards of writing on aviation topics is the interesting people you meet. Rick is one of the most unusual. I have never known anyone more consumed by a passion for flying. He is fascinated by its history and excited by its impact on modern life. He lives and breathes aviation.

As a teenager he traded aircraft paintings for flying time and soloed on his 16th birthday. He qualified as an A&E mechanic and was hired by United Airlines, meanwhile doing paintings for its pilots. In 1971 he took a gutsy gamble, giving up his job for a career in aviation art. He has been inducted into the Colorado Aviation Hall of Fame, the youngest individual to be so honored.

Artistic success in a specialized field like ours requires more than enthusiasm and talent. Painstaking research lies behind works that are imaginative and technically correct. Rick spends hours with his extensive references before picking up a brush. When possible, he personally examines and photographs a subject aircraft. Research is the dull but necessary chore that reveals vital background information. Occasionally it produces the sort of surprising story he related.

He had been commissioned by the USAF Academy to paint a mural depicting the several types of aircraft that have been used in cadet training. Academy records showed that two Aero Commander L-26Cs operated as parachute training jump aircraft for seven years beginning in 1970. Basically, the L-26C was a Model 560 modified to 660 configuration with the installation of 340-hp supercharged Lycoming engines. Interestingly, the two ships, numbered 55-4647 and 55-4648, had previously served with the 1254th Air Transport Wing at that time at National Airport, which provides presidential and other VIP transport.

Robert C. Mikesh, senior curator of the Air and Space Museum, unearthed a wealth of presidential aircraft data that was published in 1963.

From this material, Rick learned that 55-4648 had served as *Air Force One* when President Eisenhower rode it between Washington and his farm at Gettysburg. Rick would use that number on the L-26 in his painting. Further research disclosed that both academy L-26s were retired in 1977 and sent to Davis-Monthan AFB for storage. According to records, 55-4648 went to the Wedell-Williams Memorial Foundation Museum at Patterson, Louisiana, while 55-4647 was scrapped.

"I knew my old friend, Bill Duff, had several 680 Commanders in his salvage yard at Denver I could shoot for details," said Rick. "He asked about my interest in academy jump ships. When I explained, he said, 'Why don't you examine the real thing?' He led me to a Commander that has gathered dust for years, we found the nameplate and there it was, 'Serial Number 55-4647.' I couldn't believe it!"

The logs revealed that 55-4647 was not scrapped but transferred to the Nebraska Civil Air Patrol. Duff, unaware of its significance, traded engines for it and ferried it to Denver where it sat neglected for nine years. The USAF Museum at Wright-Patterson AFB hopes to add the first lightplane operated as *Air Force One* to its collection at Dayton. Indeed, it should be restored and displayed; it is part of our history.

Twelve of our presidents have flown, two as pilots. Presidential interest in flight goes back to George Washington. Along with the four men who succeeded him, he watched Frenchman Pierre Blanchard's balloon ascent at Philadelphia in 1793, the first manned flight made in this country. In 1910, two years after his term ended, Theodore Roosevelt was Arch Hoxey's passenger on a Wright Type B at St. Louis, making him the first president to fly.

Franklin Roosevelt flew from Albany, N.Y., to Atlantic City in a Ford Tri-Motor to accept the Democratic nomination in 1932, arousing worries that he might continue flying if he won the White House. At Secret Service insistence he travelled by train until World War II when the time and risk involved in sea travel made flying an acceptable option for international trips.

Roosevelt flew to the coast of west Africa in 1943 in a Boeing 314 four-engine flying boat. It was Pan American Airways' *Dixie Clipper* prior to being commandeered by the Navy. Had it operated using today's call signs, it would have reported as, *Navy One*. The remaining distance to Casablanca was flown in Army C-54s operated by ex-TWA crews. The return trip was completed without incident.

A specially-equipped C-54A was ordered for presidential use and assembled under tight security, a unique feature being an elevator that lifted the president's wheelchair into the cabin. Semiofficially called *The Flying White House*, a reporter dubbed it *The Sacred Cow*, which stuck—to the chagrin of officialdom. It was ready in time for Roosevelt's historic trip to Yalta, his third and last flying journey and the only time he used the airplane ordered for his use.

Roosevelt was not fearful of flying, but found it confining. He most enjoyed sitting between the pilots on a platform that raised his wheelchair to flight deck level. Harry Truman on the other hand relished flying and was more than once chided by the press for trips seemingly more personal than official. At his boss' request, Major Hank Myers once buzzed the White House at fewer than 500 feet, eliciting more media scolding.

The *Cow* was used by a number of other notables including Winston Churchill, Madame Chiang Kai-Shek, Gen. Sikorski of Poland, and Herbert Hoover. It flew Gen. Eisenhower home to a hero's welcome after World War II. During 1,900 hours of VIP duty it visited 44 nations and crossed 11 more.

The ready acceptance of flying by two presidents and numerous dignitaries gave commercial air travel a welcome boost. After all, the *Cow* and the *Independence*, which replaced it, were little different from stock DC-4s and -6s flown by airlines.

Builders had understandably mixed feelings about VIP aircraft. The honor of being selected for White House duty was tempered by the specter of an incident—or worse. No matter what the actual cause, or even if the president was on board, the repercussions hardly bore contemplation. Whatever apprehensions might have worried Douglas officials were relieved when President Eisenhower requested a Lockheed Constellation, the type he used as commander of SHAPE in Europe. The *Columbine II* was basically a Model 749, the later *Columbine III* an 18-foot longer 1049 of increased range and speed.

The Secret Service wanted the president in four-engine aircraft but after much discussion it was decided a small twin could safely fly him to his Gettysburg farm, thus the first use of a lightplane as *Air Force One*. Until a customized L-26 was completed, he made the 73-mile hop in an off-the-shelf demonstrator loaned by the builder. It is rumored that he enjoyed sitting in the copilot's seat. Following his heart attack, a bed was installed.

Presidents Kennedy and Johnson used the *Independence* until the first VC-137s were delivered. Now Boeing had the honor and responsibility, the sleek turbojet being a standard 707-320, except for its furnishings and confidential equipment.

President Johnson, like Eisenhower, had a country home with a runway unsuitable for the Boeing. A single Beech A90 King Air was ordered under the designation VC-6A and used to fly the president from Bergstrom AFB at Austin, Texas, to the LBJ Ranch. This commute was facetiously dubbed "The Lady Bird Special" to the First Lady's annoyance.

The historic *Cow* is presently undergoing restoration at the USAF Museum where the *Independence* and *Columbine III* are already exhibited. Also there are the Beech VC-6A, a VC-140B (Lockheed Jetstar) used by four presidents, and two presidential helicopters. Eventually a Boeing VC-137 from the VIP fleet will be added to the collection.

Each was in its time as much a national symbol as the Statue of Liberty. When the Eisenhower L-26 joins its sister *Air Force Ones*, as surely it must, I hope everyone remembers that Rick Broome saved it.

20

Not on the checklist

WE WERE CRUISING THROUGH WET SNOW AT NIGHT with airfoil, prop, and windshield heat on and were overhead the Bluefield, West Virginia, VOR when all four engines slowly died. It was one of those bizarre events for which there was no checklist or book procedure. We were not totally surprised; it had happened before on DC-7s. The engineer worked to restore power, the captain turned off the airway, and I called for a lower altitude.

Indianapolis Center said "Stand by," which was no surprise at all. We drifted down 2,000 feet before power was restored. Screens (later removed) in the scoops had iced over, restricting intake air flow. We had been at 24,000 feet and expected warmer air to melt the ice, which it did, so no big deal. But I set my tray of chicken-and-peas dinner aside until the big Wrights went back to work.

The loss of power in a single is downright exciting, particularly when low, at night or in cloud, and all the more so if you're not sitting on a parachute. There's an emergency checklist that might help set things right. The multiengine pilot can continue with an engine out, at least to the extent of stretching descent. Three- and four-engine crews can stay aloft with multiple failures. After fuel was dumped the 727 and turboprop Electra did quite well on one engine.

There's no procedure for the loss of everything in multi-engine types, but it's happened. A DC-6 freighter leaving LaGuardia Field suddenly grew silent as the gear came up. The crew ditched and survived. The ADI (water injection) system had been serviced with hydraulic fluid. A Beech D18 lost both engines after takeoff at Dallas and bellied in between houses, having been filled with jet fuel. A 767 ran tanks dry and was force-landed on an abandoned airfield in Canada. A fueling mixup and faulty gauges were blamed. And there was a 737 in Louisiana that deadsticked off airport, rolling safely on a grass embankment in ghostly silence.

The only mention I ever heard in school of total multi-engine breakdown was from a ground school instructor who said, smiling at the absurdity, that if all four quit on a 747, their windmilling would supply sufficient hydraulics to maintain control down to 160 knots. He was right; a 747 lost all four over the Pacific and was manageable until relights were accomplished. Another 747 experienced total power failure near an erupting volcano and was controllable until the fires were relit.

Nor is there a published procedure for the actual loss of an engine—that is, when one actually parts company with the airplane. There have been a number of such incidents on recips and jets, most without disastrous consequence. Recently the crew of a twin-jet airliner was stunned to learn from the tower that the ailing turbine shut down en route was missing from the wing.

Years ago an English magazine carried a picture of an intact Fleet biplane on floats without its engine that had torn loose at 3,000 feet. The copilot of an Electra—the original Lockheed twin—was alarmed when the right prop sliced into the cockpit next to his knee, the outboard engine mount bolts having sheered. He climbed into the captain's lap, the obvious "immediate action." They laughed about that—later.

In addition to engines, airplanes have shed props, sections of wing and tail, cowlings, wheels, flaps and doors, usually without dire effect. I watched a rudderless AT-6 land at Tulsa in a gusty wind. He got it on the runway, lost control and cartwheeled, but both occupants walked away. No one who saw the amazing picture can forget the B-52 successfully landed after losing its rudder and all but a stub of vertical stabilizer.

A 737 crew encountered a problem not covered on the checklist. When power was advanced for takeoff, one engine accelerated to 115 percent of allowable thrust. Retarding the thrust lever had no affect so both levers were pulled to idle. The faulty engine raced on and was killed by shutting off its fuel supply.

A DC-4 crew en route from Amarillo to Denver found the way blocked by a line of thunderstorms. They were able to slip across a "saddle" between two build-ups. Halfway through the cloud valley, riding smoothly in the clear, they ran into golf ball-sized hail pitched out by one of the cells.

The potentially lethal shower lasted eight seconds yet it knocked out the windshields, smashed the nose and leading edges, and peeled fabric from the tail. Despite the savage beating, the redoubtable -4 was safely landed at Colorado Springs. Six weeks were needed to patch it for ferrying to the factory for rebuilding.

We never ran up our P-51 Merlins past 40 inches of manifold pressure during mag check else the tail would rise even with the stick full back. The same was true of Spitfires. During max power tests two RAF mechanics would sit on the stabilizer. What happened was perhaps inevitable. Upon becoming airborne a Spit pilot experienced extreme tail heaviness and saw in his mirror a terrified mechanic clinging to the rudder, a dicey situation, particularly for the stowaway. Sharp flying around the circuit and during the ticklish approach saw the windblown lad safely back on terra firma—and if that didn't earn him a 48-hour pass it should have.

In the 1950s, Nashville offered an ILS approach to Runway 2 at Nashville and a back course to Runway 20. Of course, these approaches were supposed to be made one at a time. One morning we broke out at minimums for Runway 20 to see another DC-6 landing toward us on Runway 2. We crammed on full power, banked right, and roared across the rooftops at 100 feet.

Our unorthodox go-around was disconcerting, not dangerous, and had a humorous aspect. We caught a glimpse of a housewife hanging out wash, probably with a mouthful of clothes pins. We wondered how many she swallowed when she saw a huge airplane at full power approaching just above the trees.

If a window blows out, lost cabin pressure requires a rapid descent to a life-supporting altitude. No one thought of a windshield failing until the captain's blew out on a BAC 1-11. He would have gone completely out of the cockpit and been lost had not a cabin attendant grabbed his legs. Badly battered, he survived descent and landing by the copilot, still riding outside from the waist up.

Pilots had long been assured that modern technology detected fatigue damage long before it posed safety hazards. Whatever faith they had in that notion was shattered when the cabin roof peeled off an Aloha Airlines jet, and other high-time airplanes were found, upon closer inspection, to contain serious structural defects.

The emergency descent and engine-out procedures practiced in simulator training hardly prepared Captain Cronin for the sudden loss of the side of his 747, two engines and damage to flight controls. Nor was Captain Haynes trained to fly a DC-10 without hydraulic boosters when an engine, supposedly magnafluxed for flaws, disintegrated.

Airline history is replete with incidents wherein pilots solved unusual inflight problems. Coping with the unexpected goes with the job. More often than not, sharp piloting in potentially disastrous predicaments goes unnoticed by the media. Airlines wince at any mention of excitement; passengers don't relish thrills. Builders likewise downplay questions about their products.

Industry and government officials are forever claiming that most accidents are caused by pilot error. Depending upon which desk-bound experts you read, cockpit mistakes are to blame for 75 to 90 percent of airline accidents. In view of the accident history well known to these smug insiders, their finger pointing does not do them credit.

Many the pilot who found himself in terrible circumstances due to a design error, improper maintenance, inadequate information, or unrealistic procedures. Many the day saved when superb flying skill saw a flight safely completed in spite of mistakes made by other people. Every airliner flying has been modified to correct flaws in its original—federally approved—design. Every current procedure with regard to maintenance, training, and operations reflects lessons learned out on the line, sometimes at terrible cost—by pilots.

The flying fraternity would regard ground people with more respect and less distrust if they accepted their share of the blame when things go wrong as readily as they share the praise when things go right.

21

Complacency

THE DC-3 KEPT US BUSY. We were forever fiddling with manifold pressures and props to keep engines in sync. The elevator trim wheel's knurled rim was worn smooth by a thousand sweaty palms; the man on the controls knew the hostess was bringing up coffee before she opened the door. Oil pressures, cylinder head temps, and voltages strayed from the green often enough to require continuous monitoring.

The "Grand Old Lady" of air travel was reliable and undemanding in respectful hands but the old girl brooked few liberties and was quick to upbraid the inattentive. You paid attention or were sternly warned to sit up straight and mind your manners.

The DC-4 was bigger and faster but equally trustworthy and forgiving. It was heavier on the controls but had a rudimentary autopilot to share the workload on long legs. There was more to do and watch but an alert crew received few rude surprises. Flying the -4, a jumbo in its day, was agreeable and prestigious employment. Soon after World War II a C-54 (the military version) crossed the Atlantic without its crew once touching the controls from power advancement to roll out, a hint of things to come much later.

The DC-6 was bigger and faster still and came with pressurization, reversible pitch props, water injection, wing and tail heat, and other advances—all prone to malfunction. A good autopilot was installed and there was a flight engineer to mind the machinery while the pilots flew. The Convair was similarly updated but had no autopilot or engineer. In certain anxious situations its copilot was the busiest man on the payroll.

Flying recips during the thunderstorm months and in winter storms was character building. You didn't sit back and watch things happen. You didn't manage the airplane, you flew it and were ever alert for the sights, sounds, feels, and smells that spelled good things or trouble. It was hands-on work that now and then called up every trick you knew and, in rare events, ideas you hadn't tried before. There was nothing boring about any of it, yet those times were tremendous fun so long as you stayed ahead of the airplane.

The introduction of turbine power brought dramatic changes. The 707 was two-and-a-half times heavier and 250 miles an hour faster than the piston equipment it replaced. At first blush it seemed hopelessly complex yet the transition to its awesome performance proved relatively easy. Operation of the turbojet engine was simplicity itself—shove thrust levers forward to go, retard them to descend—and it almost never posed problems. Many pilots went 20 years without a single breakdown. Compared with even the most reliable piston-prop power combinations, the turbines seemed too good to be true.

The 707's complicated flight controls, hydraulic and electrical systems, and advanced radio and nav gear were nearly as trouble-free. Automatic devices actually functioned on their own without prodding; you could depend on them. On long trip legs with the sleek jets smoothly humming and the autopilot tracking airway radials with uncanny precision, the resulting tranquil cockpit atmosphere tended to bore.

Manually controlling a large transport with the precision expected in airline flying requires intense concentration; monitoring its automated flight is child's play. We were vaguely aware of an insidious threat creeping into the picture: complacency. It would prove a threat as lethal as a terrorist's bomb.

The latest generation of airliners and business jets embodies stunning technological advance in two respects. First, the airplane, its engines, and systems have been made reliable to a degree undreamed of in the piston era. They are virtually foolproof. Secondly, the airplane has been made almost totally self-operating.

It can be programmed to climb, cruise, descend, hold, land, and brake untouched by pilot hands. Set the dials, type in the numbers, and it will fly far more precisely and economically than is possible under human control. Indeed, the precise and fuel efficient flight now possible depend on the continuous use of its automated systems.

Now comes "fly-by-wire," clearly the wave of the future. The FBW pilot does not physically manipulate flight controls or adjust power, in fact, there are no mechanical linkages between cockpit and flight controls or engines. Movements of the stick or wheel and thrust levers send signals to computers that activate hydraulic boosters and fuel controllers. These black boxes "remember" airframe and powerplant limits and will not allow a pilot to exceed them. Compared with the military and civil FBW aircraft now in service, the first jet types we thought so marvelous were quite primitive.

Further automation is logical, indeed inevitable. As a hint of what's coming, consider that the radical B-2 bomber's systems are configured for takeoff with a single switch, another puts it in landing mode, and a third readies it for its "stealthy" combat role. The day of the push-button pilot at hand. Of course, he will be trained to assume control in case of a malfunction, but for the most part he will sit and watch as computers do a job he could never match.

All of which is a mixed blessing. In reducing human involvement to a minimum, designers unwittingly have increased the threat of pilot laxity and mistakes. A demanding airplane keeps a pilot on his toes; tomorrow's self-flying airplane cannot avoid inducing boredom that can lead to complacency. The pilots most likely to be affected are the crews of medium- to long-range airline trips because of the routine nature of their work. I've known pilots who flew the same schedules in the same equipment for five years. After flying the same run month after month without an inkling of trouble, it is only natural to relax and subconsciously shrug off unpleasant possibilities.

What's the solution? A fresh, hard look at the human side of the man/machine equation is needed. From the beginning the theme has been to adapt pilot to airplane instead of doing it the other way around. Traditionally, the pilot has been factored into the equation as a constant instead of a variable; this is unrealistic. A hydraulic pump will (nearly every time) function as well at the end of a long hop as it did when the engines were started. The pilot who completes the same trip is not the same fellow who reported for duty 12 hours earlier. The pump is a near-constant, the pilot a variable.

Human performance is adversely affected by fatigue, distraction, frustration, concern, health, and other reasons. No two pilots perform in exactly the same way in every situation, no matter how standardized their training. Neither does an individual respond with the same proficiency every time, no matter how conscientious he might be.

The finest engine or component will fail sooner or later, even if but once in a million times. Designers recognize these certainties and provide redundancy. The modern aircraft is as "fail safe" as they can make it. Similar allowances are not made for pilots who are equally certain to fail, again if only in the millionth case.

"Human engineering" has been paid much lip service but given too little serious attention. Warning lights, horns, taped admonitions, and stick shakers have not prevented accidents that should never have happened. Designers would profit from riding in the cockpits of their marvelous creations, not on factory test hops, but out in the real world of frustrating traffic congestion, mean weather, long night hours, and slick runways. They should sense firsthand the monotony and fatigue that pervade the pilot's everyday environment and make it all too easy for him to commit a terrible blunder.

Complacency, carelessness, and inattention are not always examples of unforgivable behavior. All humans are prone to occasional, if rare, lapses of attention and good judgment. Allowances have not been made for them. A shameful pilot error verdict often ignores a design concept needing reappraisal. If one percent of the cost of designing the next airplane was invested in exploring the true nature of the people who will operate it, safety would surely be enhanced.

22

The age-60 rule

WHEN CAPTAIN DAVE CRONIN LANDED at Honolulu February, 1989, his overweight 747 had a 10-foot by 20-foot hole in its side, both right engines shut down, and flaps that would only extend 10-degrees. He put it on the ground at 195 knots, touching down "just like a normal landing," according to one passenger. The *Wall Street Journal* headlined, "United's Flight 811 Showed How Vital Capable Pilots Can Be." Cronin, 59, had one trip left to fly before retirement.

Five months later, UAL's Capt. Al Haynes somehow kept his DC-10 flying for 41 minutes with its controls set in concrete, having lost all hydraulic boosters. The chances of survival for anyone on board were virtually nil, yet Haynes and his crew (he and Cronin would reject the inference either pulled it off alone) managed an arrival at Sioux City that saw nearly two thirds of the souls aboard survive. "What if a less skilled pilot had been at the controls?" said TIME. Haynes would retire in two years.

Why are pilots of such outstanding caliber grounded at age 60? It's the law. Specifically, it is Federal Aviation Regulations Part 121 that states, in part, "No person may serve as a pilot on an airplane engaged in operations under this part if that person has reached his 60th birthday." The background is interesting.

President Eisenhower appointed his friend, ex-USAF General Elwood "Pete" Quesada to head the FAA in 1958. Airline pilots were not Pete's favorite people, in fact, he held them responsible for most safety problems and vowed to bring airline flying "up to military standards." Under Quesada, the FAA, which theretofore had shown no concern about pilot age, proposed two new restrictions, one prohibiting any pilot over 55 from qualifying on turbojet aircraft, the other grounding all pilots at age 60.

After a public hearing the age 55 proposal was dropped. Despite a deluge of objections by industry and medical experts, the "age-60 rule" was implemented in 1960—without a public hearing. Senior pilots with impeccable records found themselves on the street. (A suggestion that social security payments should commence immediately because employment had been terminated by federal edict fell on deaf ears in Congress.) All remaining pilots faced the loss of their five most lucrative years.

Air Line Pilots Association president Charley Ruby protested but received little membership support. The majority of younger members favored the rule; now they would move up in seniority faster. It was an unthinking, self-serving stance unworthy of them but the egoistic get-mine-while-I-can syndrome was irresistible.

J.J. O'Donnell, Ruby's successor, half-heartedly continued the fight but eventually ALPA bowed to the consensus and thereafter sided with the FAA. No one knew better than ALPA's leadership that the age-60 rule was wrong, yet assuaging the membership took precedence. Senior card holders who had built the union and who paid the highest dues were summarily abandoned. It was not ALPA's finest hour.

A number of campaigns were mounted to overturn the rules, mainly financed by older pilots and retirees. All failed. At one point Congress ordered the National Institute on Aging to conduct an in-depth study of the matter. After three years of deliberations and the expenditure of $250 million of taxpayer money, that august body found "no medical significance to age 60 as a mandatory age for retirement." However, it recommended that the rule be retained because "age-related changes in medical conditions and performances" would have decremental effect on some individuals.

The report ignored certain facts and made a mockery of scientific research. At least the panel urged the FAA to allow selected volunteers to continue flying past 60 under close surveillance in order to acquire the age/performance data it claimed were unavailable. This reasonable proposal was ignored. Six years later several retirees instigated a court-ordered review of the rule. The FAA's response to this latest attempt: exemption denied.

The FAA maintains that older pilots are more likely to be involved in accidents. It acknowledges that there would be individual exceptions but points out that the "population as a whole" is subject to an increased rate of disability with increasing age. This is hardly news. Neither is it news that we don't all fit the norm; we age at different rates. I knew pilots grounded in their 30s and 40s. A Mayo Clinic study in 1968 predicted that about 20 percent of all airline pilots would be medically disqualified before they reached retirement age.

Yet, as a senior copilot I flew with captains nearing 60 who were as sharp as men 20 years their junior. More than once we shared predicaments requiring heads-up piloting and they performed faultlessly in every instance. To judge such individuals by a general pattern of physical and mental deterioration is unrealistic and grossly unfair.

According to the FAA, there is no way to discriminate between those over-60 pilots who could still safely fly and those who could not. The excuse that such data are unavailable is weak. The age-60 rule did not ground every pilot.

After 1960 a number of retirees continued flying airline equipment in charter work or with overseas carriers. How have they fared? And, had NIA's suggestion been followed, by now the FAA would have years of firsthand data to support (or refute) its conclusions. Apparently there is reluctance to acquire new data that might upset preconceived notions.

Petitioners for modification of the age-60 rule argued that experience more than offsets the risks of sudden incapacitation brought on by aging. ALPA countered that after 5,000 hours of flying, "additional flight time does not significantly improve pilot performance from a safety standpoint." The FAA agreed. It also accepted ALPA's claim that "experience does not prepare a pilot for technological change," an absurd contention. I have friends who began airline flying in 90 mph Fords and ended their careers in 600 mph Boeings.

The incredible flying of Cronin and Haynes and the expert handling of many other emergency situations by older pilots do not unduly impress the FAA. "What the pilot of UAL Flight 811 did would be expected of any captain," it said of Cronin, adding that it "cannot base its decision on isolated commendable acts."

The FAA's refusal to grant an exemption to the age-60 rule or allow a closely monitored trial, is inconsistent with its willingness to recertify under-60 pilots with serious problems. Pilots with histories of alcoholism, myocardial infarction, bypass surgery, monocular vision, personality disorders, strokes and dysrhythmias have been returned to flight status. During a nine-year period, more than 400 pilots rehabilitated from alcoholism were returned to airline flying even though the relapse rate was known to be 20 percent.

Statistically, the pilot who has had a heart attack or stroke, bypass surgery, a drinking problem, or lost an eye has a 35 percent chance of regaining his first-class medical certificate. That means he can resume airline flying providing, of course, he's less than 60. "The feds say I became a poor risk at age 60. How could they be sure I was safe at age 59?" said a friend who flew for 35 years without taking a day's sick leave. He's still hale and hearty 14 years later, by the way. Did the physical examinations and proficiency checks that kept tabs on fitness and performance throughout his career become less revealing on his 60th birthday?

The age-60 rule will ground 5,000 pilots during the next five years. That it remains on the books nearly three decades later shames all younger pilots who support it and the bureaucrats who stubbornly refuse to admit a mistake. It should be abolished. In the meantime the flying public would be better served if the FAA worried less about the pilots it certifies and more about the airliners it approves. The recent rash of safety problems was caused by older airplanes bearing the FAA's stamp of approval, not by older pilots.

23

Top Gun talk

THE SHOW WAS BILLED AS A TELEVISION DOCUMENTARY on fighter pilots, but turned out to be more of a tribute to them, with most of the tribute coming from the "top guns" themselves. The Mach 2 jocks talked about "living on the edge," "pushing the outside of the flight envelope," and other death-defying aspects of their trade in self-laudatory terms. Modesty has never been overworked in fighter circles and their hazardous work certainly excuses a bit of good humored braggadocio. If only they'd left it at that—but no, they had to put down the rest of us.

"Flying an airliner is like driving a bus; you couldn't pay me to do that," said one. Another talked about discipline in a way suggesting that nonguns don't know what the word means. A third said, "Flying for the airlines is as exciting as watching a fellow get a haircut." You have to wonder about people who promote themselves by deriding others.

There's nothing new about any of it. The air-minded boys of my generation yearned to join the Army or Navy and fly "pursuit ships." We caught the bug from reading pulps like *G-8 And His Battle Aces* and watching silent epics like *Wings* with a piano player providing sound effects. The Spads and Nieuports tumbling across the silver screen to nail the Red Baron's Fokkers held us spellbound; who wanted to fly a plodding bomber?

Then came the handsome Boeing F4B-4 and curvaceous Curtiss P-6E of the 1930s and we were further convinced that pursuit pilots had it all. Some of my pals got their wish in World War II and flew Mustangs and Corsairs, by which time pursuit ships were called fighters, while the rest of us flew bombers, transports, or trainers.

We were disappointed. Fighter pilots were the adored heroes, just as they had been in the first war. Rarely did a bomber crewman receive the individual attention accorded scores of aces. Churchill eulogized "the few" Hurricane and Spitfire pilots who defended England, and indeed they deserved his

The North American F-86 earned fame in the Korean War.

praise, but it might have been noted that the RAF Bomber Command suffered heavier losses while the Battle of Britain was waged.

Prestige-wise, those of us assigned to transports would have been at the bottom of the stack but for Training Command. We were Air Transport Command. We hauled men and equipment to all theaters of operations and ferried everything from PI-23s to B-29s to wherever they were needed, including single-engine fighters across the Atlantic. Combat airmen regarded us with disdain. We were ATC, the "Army of Terrified Copilots" and "Allergic To Combat."

Our work was far less perilous than tangling with Focke-Wulf 190s or boring through flak in a B-17. Usually, that is. On the treacherous Himalayan "Hump" route between India and China, ATC losses exceeded those of combat units in the area. The weather was often appalling yet clear skies were a mixed blessing as they made unarmed transports the prey of Japanese fighters. The pilots of Troop Carrier Command likewise sustained heavy losses on many paratroop and glider towing missions. They were by definition noncombatants, meaning they were eligible targets without the means to defend themselves.

The second-string aura stuck like glue. We were "trash haulers." By the same token, bomber pilots must have been dump truck drivers, though the fighter crowd regarded them with admiration and pity. The odds of surviving 30 missions in a B-24 were not encouraging. As for the poor fellows stuck in Training Command, at least they supplied pilots to fly the fighters, bombers, and transports—and at more risk than was generally acknowledged.

It was not the adulation of fighter pilots that bothered us transport types. They got the best press, always had and always would. That went without saying. Anyway, the people we most wanted to impress were girls and they couldn't tell a P-47 pilot from an AT-6 instructor. However, we did think our participation was significant and deserved rather more credit than it received.

After the war most airline berths went to vets who had flown transports or bombers. Some fighter pilots keen on flying careers decided that bus driving or being the "glorified chauffeur" of a business aircraft might not be so bad after all. But most snorted at the idea and took up nonflying vocations, a decision many came to regret.

Postwar predictions that missiles would soon replace manned aircraft proved fanciful. We are still producing fighters by the gross and air-minded boys to fly them. They read about the great aces of past wars, watch Tom Cruise roll his Tomcat, hear fighter pilots talk about pushing the limits, and catch fighter fever.

Once strapped into their lovely jets, they pick up the old refrain, one chorus of which has it that transport pilots are bus drivers. It's part of flying lore, repeated until it becomes the gospel truth. If the fighter jock is someday displaced by a soulless rocket, we will have lost something. His gutsy, boisterous, cocky breed has written tremendous air history.

But he can be tiresome. There's nothing wrong with his believing he's the warmest throttle bender since Lindbergh, but when he puts down the rest of us he's a bore. He advertises an appalling ignorance of piloting in its broadest sense. The hotshot who claims you couldn't pay him to fly for an airline might not qualify if he applied. The transition has proved beyond more than a few applicants with his background.

He could master the technical skills, but doing things the airline way requires an attitude adjustment that he might not be able, or willing, to make. The aggressive loner who craves "living on the edge" and "pushing the outside of the flight envelope" cannot always adjust to staying well within the envelope and as far from the edge as possible. He wouldn't be his own man for years and might find the lengthy apprenticeship more than he could endure.

The mental approach that made him an F-16 ace is not quite what the airlines want. Airline flying requires strict adherence to established rules and procedures, cooperation with fellow crewmen, patience, calm thinking under stress, and endurance. An airline pilot must often perform at his best in the final minutes of a fatiguing 10-hour flight, or 14-hour working day. And he must maintain during his entire career the expertise he mastered at its beginning. Airline flying is a routine, monotonous, frustrating, and fatiguing exercise much of the time. But, for the individual who wants a lifetime of flying, it can be rewarding.

Discipline is not exclusively a fighter pilot attribute. The self control and good sense to do the right thing in every situation are characteristics of every good pilot, no matter what he flies. The urge to excel that drives a fighter pilot to push the limits of himself and his aircraft also compels an airline pilot to remain strictly within the limits of his. It is not the relative risks involved that matters, but doing both jobs well.

It is true that airline flying is normally unexciting, even dull. Pilots work hard to deliver serene air travel. The thrill-a-minute, gut-wrenching maneuvers that boost the fighter pilot's adrenaline flow are far removed from the lumbering progress of a 747 where things usually go so well that boredom is the problem, complacency the danger.

And it's not like watching haircuts every time. There will always be those proverbial moments of stark terror when weather, mechanical breakdown, fire, structural failure, fuel shortage, or some evil combination of them transforms normal into nightmare. There's no punching out when it happens; airliners don't come with ejection seats. You sit right there and sweat out a solution while 100 or 400 souls in back depend on your finding one that works.

The adventurous life of today's fighter pilot is something we want to hear about. We admire his courage, skill, and enthusiasm. He's a delightful fellow so long as he talks about the only kind of flying he knows much about.

24

Postcard from Framlingham

A FITFUL, BONE-CHILLING WIND stirred the grass and drove a persistent cold drizzle against the squat, drab two-story building. Percy Kindred looked out across the broad green fields of his Suffolk farm and saw it as it was when the Americans were there 40-odd years prior. "Yes, I was right here the whole time," he said, remembering the dawn rumblings of 100 Wright radials on mission days.

Air Ministry officials commandeered the land in 1942 and told his father the annual rent would be 15 shillings an acre. "We were only too pleased to do what we could to aid the war effort," said Kindred, adding that a few wealthy landowners complained that such use of their land would disturb the pheasants and partridges. We had some of that sort on our side, I said. Contractors worked through the winter in miserable weather. "I still marvel that it was completed," he said.

By May, 1945, when Germany surrendered, there were more than 700 service airfields in the United Kingdom, 500 of them built since the war began. Station 153, one of 112 fields allotted to the 8th Air Force, USAAF, was built on the Kindred farm. It was popularly known as Framlingham, the name of a nearby town, a name indelibly etched in the memories of those who were there.

It was nearly completed when a B-17, limping home from a raid on Kassel, appeared low over the trees, both inboards feathered. Red flares were fired, the signal that wounded were aboard. A construction truck drove onto the runway, its driver oblivious to the approaching cripple. The desperate pilot attempted to go around for another approach. As Kindred watched, the Fort slowly lifted, rolled inverted, and dove into the ground. Ten young men. Framlingham's toll had begun.

The 95th Bombardment Group arrived in May and flew several missions before taking up permanent residence at nearby Horham. Kindred remembers one of those raids. "About 18 of their B-17s took off on a mission to Germany.

Two aborted. Six of the rest came home. Of those that survived, one crash landed and the rest were literally shot to pieces. That was one of my first sights—virtually outside my front door—of what those men were going through. I was stunned. What I saw that Sunday evening I won't forget . . . ever."

What Kindred witnessed were the remnants of a force badly mauled by fighters over Kiel. Five more of the 95th's aircraft went missing that month, the highest loss sustained by any bomb group at the time.

The 390th Bomb Group, likewise a B-17 outfit, reached Framlingham via Iceland and Scotland in July and flew its first mission a month later. Its record during the following 21 months was typical of 8th Air Force heavy bombardment groups. Its B-17Fs flew 294 combat missions, dropped more than 19,000 tons of bombs, and claimed 342 enemy aircraft destroyed, 74 probables, and 97 damaged. Following VE Day it flew six food missions, carrying 436 tons of supplies.

During its brief life, Station 153's population grew from 2,170 airmen and 40 Fortresses to 3,104 men and 65 aircraft. At the end it counted 144 aircraft missing in action plus 32 other operational losses. It had won a Distinguished Unit Citation for its part in bombing Regensburg and a second for a raid on Schweinfurt. Its claim of enemy aircraft destroyed on a single mission surpassed that of any other group.

So much for the bare facts. Behind them were uncounted—and often unknown—tales of fear, heroism, triumph, and tragedy. While some bomber

Spitfire cockpit.

crewmen survived enemy action to become prisoners of war, most did not. More than 1,000 American airmen were killed flying from Framlingham, the most telling statistic of all.

The 390th's "other operational losses" made up 18 percent of the total number of aircraft destroyed or damaged beyond repair. This percentage was typical among bomb groups based in Britain. A number of reasons can be cited. First, the level of pilot experience was relatively low considering the complexity and reliability of multi-engine aircraft of that day, coupled with the extreme demands of combat flying. Operations were often conducted in appalling weather with minimum navigational aids. Weary and sometimes injured young pilots were often hard-pressed to keep their damaged ships flying, much less to pull off safe landings.

The concentrated firepower that was the best defense against enemy fighters required close formation flying, which increased the risk of collisions. One account reports pieces of an exploding B-17 disabling the bomber on its wing, which climbed into the one above it. Five chutes were counted as the three ships and 30 crewmen fell to earth. There were many collisions in cloud and more near the bomber bases in England.

Most of the 500 airfields built in Britain during the war were clustered in the south of England in an area roughly the size of South Carolina. Enough concrete was poured in their construction to build a six-lane highway from New York to Orlando. The concentration of fields in the southeast region known as East Angelia was such that traffic patterns almost overlapped. The problems involved in assembling hundreds of bombers from scores of bases into a unified disciplined force were equalled by those of recovering weary and battered crews hours later.

Mechanics, most of whom had never touched an airplane two years earlier, worked outside through the night often in bone-chilling weather, completing routine checks and repairing battle damage. That the daylight precision bombing plan succeeded as well as it did seems incredible from this distance; that the losses were not far greater is a tribute to everyone involved.

Then it was over and the bustling air bases were suddenly silent. Within weeks the bombers flew away for the last time; the ground people followed soon after. Framlingham reverted to its owners and the runways, taxiways, and revetments were broken up and removed. Farm land in Britain is too precious to leave under concrete. Today you can fly across most of the wartime airfields and never guess they were there. The few hangars and quonset huts remaining are used to store farm machinery.

The drab two-story building with the outside balcony and glassed cupola was of course the control tower. With Percy Kindred's cooperation it has been painstakingly restored by air history enthusiasts led by Ron Buxton. It now houses the 390th Bomb Group Memorial Air Museum. The hard work required is obvious as was my question, "Considering the record of your own Bomber Command, why all this at an American base?"

"We lived here in East Angelia near the U.S. bases," said Buxton. "We remember the Forts and Libs and Mustangs. Most of us were schoolboys then. It's history that should be preserved."

Inside are grim exhibits: mangled prop blades, oxygen tanks, a turret, smashed engines, a gear strut from a shot-up P-51 that almost made it, dog tags found eight feet down. There are paintings, photographs, letters, bits and pieces of the past donated by men who were there. A number of former 390th members attended the dedication in 1981.

If this sort of thing interests you, by all means go, but first read *The Mighty Eighth* and *Airfields of the Eighth* by Roger Freeman and Ian Hawkins' *Munster: The Way It Was*, a superlative account of perhaps the most savagely opposed raid flown by the 8th Air Force.

Percy Kindred looked across his fields, perhaps seeing again the line of nose-to-tail Forts moving toward the runway that October morning. *Miss Behavin'*, *Spider*, *Cash & Carrie*, and *Tech Supply* were in that line—and among the seven missing in action by evening. Thirty 8th Air Force bombers and a fighter were lost on the Munster mission.

"Yes, I was here," he said. "I was here the whole time." I wanted to hear more but he said nothing else. We went down the steel steps and back inside to get out of the chilling rain.

25

The first air war

ON THE 11TH HOUR OF THE 11TH DAY OF THE 11TH MONTH OF 1918, the "Great War" ended. History does not record a more savage or senseless conflict. The horrors of trench warfare are beyond imagination, the carnage almost beyond belief. French and German casualties during the 10-month Battle of Verdun equalled the casualties of all the British Empire during World War II. Then, on the first day of the Battle of the Somme, the British Army suffered 60,000 casualties, 15 times its losses on D-Day 28 years later. The appalling slaughter lasted more than four years.

"Armistice Day," November 11 (later Veterans Day in this country), commemorated its conclusion. The slaughter that ended late in 1918 is now called World War I. It was also the first air war.

The airplane was 10 years old when it was drafted. The fragile toy of daring sportsmen, it was maddening to keep airworthy and perilous to fly. An unlikely weapon, there was one thing it could do: carry an observer beyond front lines and bring him back. As the Duke of Wellington said a century before, victory belongs to the general who can guess what's happening on the far side of the hill.

The recruit soon proved itself. A Royal Flying Corps pilot spotted a column moving away from the front. The French line had broken and was retreating. British elements were quickly withdrawn, cheating the enemy of a chance to encircle thousands of troops; henceforth, there were rarely enough airplanes for reconnaissance. There would be no more guessing.

Military airmen sought more ways to make their aircraft true weapons. Dispatchs from the rear were dropped on forward positions; artillerymen, watching purposeful maneuvering overhead, laid shells on unseen targets; infantry officers were flown across the lines to look for themselves.

Inevitably, pilots began dropping small bombs onto troop concentrations. One device was the French *fléchette*, a wicked little steel dart. Tossed overboard by the gross, its effectiveness depended on a direct hit. One struck a cavalryman, passing through his steel helmet, his body, and his mount.

Unarmed Allied airmen roamed at will above German territory, their main concern being engine failure. The enemy enjoyed the same immunity. Opposing pilots waved at each other; neither had the means to harm the other. It couldn't last. Obviously, the advantage lay with the side enjoying exclusive use the air. Airmen began exchanging rifle fire and throwing grenades at each other, rarely with success.

The "scout" emerged, a single-seater with machine guns synchronized to fire between propeller blades. Able to stalk their prey and force a fight, scout pilots developed tactics basically unchanged in later wars. They captured the fancy of a public sickened by the mass butchery of trench warfare. Lines of soldiers walking into murderous machine gun fire was one thing; lofty dueling by these bold "knights of the air" was something else. Air fighting seemed a romantic affair.

Star players of the deadly new game became household names. They were the "aces." Every English schoolboy read about Albert Ball who claimed 44 victories and Billy Bishop from Canada who got 72. And Roy Collishaw of the legendary "Black Flight" with 60 and Mick Mannock who, despite poor eyesight, topped the list with 73. Sixty-five British pilots were credited with 20 or more victories.

Slight, ailing Georges Guynemer was a French national hero with 54 victories when he vanished. No trace was found of him or his plane. Flamboyant Charles Nungesser got 54, only to perish in an attempt to fly the Atlantic in 1927. Rene Fonck led the French with 75.

The end of the war was but 14 months off when the first American air unit reached France, yet former race driver Eddie Rickenbacker ran up a score of 26 before it was over. Legendary Frank Luke of Arizona, the "Balloon Buster," downed 21 aircraft in as many days at the front, among them many observation balloons, the most dangerous of targets.

The highest score (80) was claimed by German idol Manfred von Richthofen, the redoubtable "Red Baron." His brother, Lothar, shot down 40. Ernst Udet (62) and Hermann Goering (22) would be heard from again in World War II.

The accolade, "ace," was not officially recognized but popularly awarded a pilot who had downed 10 (later five) enemy aircraft. He was front page news, often pictured with top brass and government leaders. The same adulation of aces would be heaped upon top fighter pilots in later wars. The brave crews who flew observation and bombing missions in slow, vulnerable two-seaters drew little individual attention and neither would their sons in Lancasters and B-17s, though their duties were no less hazardous.

The lot of a World War I airman was far from glamorous. While he lived behind the lines in relative comfort and safety, the odds of survival were discouraging. A green scout pilot was easy prey for an enemy veteran and his replacement sometimes reported with 25 hours total time. The average life span in some hard-pressed squadrons was three weeks.

Some of the airplanes were difficult to fly and many were prone to structural failure. A single bullet in the fuel tank could set them afire, in which event the choice was often to jump rather than burn, there being no parachutes. It was no game for the timid.

Technical progress during World War I was astounding. The faltering sportplanes mustered in the beginning were succeeded by true fighting machines. By the Armistice, Britain had a souped-up scout in production that scrambled at 1,800 feet a minute and could do 150 mph. Huge biplane bombers able to carry 7,500 pounds of bombs were under construction.

The first war in the air inspired thousands of books, a slew of postwar pulp magazines and a number of screen epics. Much of what was written and filmed was lurid nonsense but it fired the imagination of air-minded boys in the 1920s and -30s, often being the only contact they had with flying. "You sent us 25,000 cadets," General Arnold, World War II chief of the USAAF, told one ace-turned-writer.

The subject intrigues modern historians who continue to unearth old records and photographs that flesh out an authentic account of that remarkable era. The League of World War I Aviation Historians held a convention at the University of Texas at Dallas, home for the History of Aviation Collection, a massive holding of aeronautical books and artifacts. [The books and other research material remain at the university; however, the artifacts have been moved to exhibit space at the main terminal at Love Field in Dallas.]

Present at the Texas meeting were George Vaughn, Doug Campbell, and Roy Brooks, who between them scored 25 victories in that distant war. Campbell was the first American ace, Vaughn the fourth ranking with 13. Brook's Spad is preserved at the Smithsonian. The sprightly nonagenarians reminisced, their memories clear.

"It was the most exciting adventure of all time . . . the Camel was maneuverable but difficult to fly . . . the S.E.5 climbed faster, comfortable, very nice . . . without question, the best German airplane was the Fokker D-7 . . . we had no parachutes . . . the real heroes were those who forced themselves into the cockpit, knowing the odds . . . I got 50 hours training, some others much less . . . 50 hours between engine overhauls . . . our success removed some doubts about air power and pointed the way . . . ."

They remembered training in Texas, living in tents, troopships to Liverpool, French trains, leaves in Paris, the colorful characters they knew, humorous experiences. Typically, they talked around the reason for it all. There is a point beyond which such warriors rarely go.

W.E. Johns, an RFC pilot, went this far: "I went back to the old battlefields. Only the names are left—Le Cateau, Douai, et cetera. No signs of war except the interminable cemeteries. All I could think was, what a waste it all was."

26

Harvard 3064

"A TRAINER YOU FLEW AS A STUDENT in the Royal Canadian Air Force is undergoing restoration here." Martin Pengelly, a *FLYING* reader in England had written me. "You knew it then as Harvard No. 3064."

He's right; it's here in the treasured old log.

How he knew that and why I recall that particular entry is a story spanning a half-century.

During World War II, U.S.-built training airplanes in British Empire air forces were named after American universities. Thus the North American AT-6/SNJ (its U.S. Army/Navy designations) became the "Harvard" at Empire schools. A low-wing design powered by a nine-cylinder Pratt & Whitney engine, it was the most widely used Allied advanced trainer then and for years thereafter; in fact, the RCAF retired its Harvards 25 years after the war. Like the famous DC-3, it was the right airplane at the right time. Happily, a good number are still flying.

Why Americans learned to fly in Canada months before their own country declared war needs explaining. Until the Pearl Harbor attack, two years of college was a prerequisite for Army or Navy pilot training. Word got around that the Canadians would accept a high school diploma. Attach no soldier of fortune aura to those who took note, for some 8,000 of us grabbed the chance. Enlistment didn't cost our citizenship; instead of swearing allegiance to the Crown, we agreed to serve for the duration, a stratagem enjoying tacit White House approval.

The RCAF curriculum was completed in three stages. At Initial Training School it was soon determined if you knew enough math and physics to absorb ground schooling. But first you learned to march. In every military service, allied and enemy, you learned to march before you learned to peel potatoes, that expertise being taught later. ITS survivors were posted to an Elementary Flying Training School to learn the basics in a Tiger Moth or Fleet Finch. My group went to No. 9 EFTS at St. Catharines, Ontario, near Niagara Falls. At "St. Kitts" we logged 65 hours in Fleets.

Graduates advanced to a Service Flying Training School for multi-engine training in Avro Ansons, or prefighter training in Harvards. I was among those sent to No. 14 SFTS, a Harvard school at Aylmer, Ontario. Of the 12 Harvards in which I logged the next 100 hours, two remain vivid in memory. During the third lesson I ground looped No. 3065.

Did I ever. It was all too easy in those early models with free-castoring tail wheels. "I've got it!" yelled Sergeant Campbell from the back seat, but it was too late. Around we wrenched on the left gear through 360 degrees before flopping back level in a cloud of dust. Amazingly, no damage was done. The Harvard was designed with ham-fisted tyros in mind.

Equally amazing, two months later I passed my "wings test," the final flight review of student progress. I sweated out many checkrides in later years, but none more than that one. It decided everything. Then my instructor pointed to No. 3064 and said, "Go fly, do anything you like." Happiness for this teenager was no more lectures, no more flight checks, no more doubts, and a glorious hour alone behind a big roaring 600-hp radial with all of the crisp winter sky for a playground. Will I ever forget November 13, 1941? Of course not.

After going overseas many of us Americans transferred to the U.S. Army and flew operational types. Every base had two or three AT-6s on hand for administrative duties. You could borrow one now and then for an hour of fun. To continue military flying after the war you joined the reserve, which paid nothing except an hour of weekend flying in an AT-6, or the Air National Guard, which wouldn't let you fly its lovely P-51s or P-47s until you'd displayed your skills—in an AT-6, of course.

My first postwar boss was Sid Park, a pioneering Kentucky pilot whose niche was aerial photogrammetry, the complex process of making surveys and maps from photographs. Before the war he mapped with Cessna C-34 Airmasters. In 1946 he reentered business with surplus AT-6s, installing a floor window under the rear seat for the large cameras he designed and built. No engine modifications were made. Our federal contracts stipulated a minimum altitude of 20,000 feet for all shots.

With pilot and cameraman aboard, plus extra rolls of film and oxygen tanks, we could reach that level in 45 minutes and remain there so long as I flew with a watchmaker's touch. Once, with the tanks almost dry, we inched up to 25,000 feet, something the designers never envisioned. My admiration for the stout trainer increased with every flight. Then an airline job made Harvard flying a pleasant memory.

I wrote a short book in 1965, *The AT-6 Harvard*, which included a reprint of the USAF pilot's manual for the T-6 (as the AT-6 was redesignated). Artist Richard Groh contributed a painting of Harvard No. 3064 rolling above the Ontario landscape. The original art hangs in our den.

Martin Pengelly in England saw the book and thought I'd like to know that an airplane I flew long ago is alive and well.

I went back to Aylmer in 1968. The few remaining barracks housed a police academy, the three wood hangars were empty and weeds grew from cracks in the runways. I half expected to find the skids marks of my groundloop 27 years earlier. It was eerie, something like the opening scene of "Twelve O'Clock High," except that No. 14's Harvards nearly always returned from their missions.

By coincidence a remarkable auction was being held at nearby Tillsonburg. In a field not visible from the road was an unbelievable array of 400 rusting vehicles and airplanes, the collection of an eccentric recluse who for years had allowed no one near. He left no will so the government was selling everything. The 50 airplanes were Swordfish and Yales.

The Swordfish was a monstrous biplane torpedo bomber with an incredible combat record despite its 100 mph speed. The Yale was the Harvard's predecessor. Similar in appearance, it had a fixed gear and 450-hp engine. The yellow paint was faded but roundels and buzz numbers were clearly visible. Wings, engines and propellers lay on the ground between the relics. Foolishly, I did not bid on anything.

One of the Yales ended up in Florida. Ray Kirkpatrick, a fellow Braniff pilot, found it to be restorable and trucked it to Texas. He asked my help with the colors. All RCAF trainers were yellow, but what shade? My snapshots are black-and-white. I put him in touch with history-minded Canadians who supplied paint swatches and plans. After five years of hard work, Ray flew his prize. You should see it. It's perfect, down to the last rivet.

A recent update from Martin includes snapshots of 3064 restored as a Navy SNJ. It now performs at air shows in formation with three other Harvards/AT-6s/SNJs. He's compiling a detailed history of its service life, which he promises to send. He says its new owner wants to put it back in its original RCAF colors as No. 3064; do I recall paint scheme details? I called Ray who dug up the names of those who helped with his Yale and I airmailed them to England.

We Aylmer students were divided into three flights. There was much rivalry between us for the fewest bent props, smashed wing tips, wheels up arrivals, and so on. Infractions required contributions to the "Rumble Fund," ranging from 25 cents for taxiing in with flaps down to $5 for a belly landing. (My groundloop cost $2.) Although my flight had the best record, there was enough in the fund to finance a memorable party at graduation.

One minor point: The Harvards assigned to each flight bore distinctive hub caps. If Harvard 3064 is to be properly marked, it must have the correct hub logo. It's a long shot, but does a reader of this book know the design used by C Flight at Aylmer in 1941? Strange things happen in the world of aging pilots and restored airplanes. I'll watch the mail.

27

Before we fly

A CAR DREW UP BESIDE THE AIRPLANE and two couples got out. While one man loaded suitcases, the other undid the ropes, and they all climbed in. They fired up and taxied out. After a quick runup, they roared off at full throttle, using a considerable distance to reach flying speed. The gear doors closed less than three minutes after the cabin door. The foursome had come in four days earlier, the gas boy said. He had filled the tanks when they arrived.

This sparking turbocharged, pressurized marvel could whisk its owner and his chums to pleasure places at close to four miles a minute, well above 20,000 feet. Its panel featured a lovely stack of avionics; type in the numbers and it would take you there and find the runway, hands off. Sit back and watch it fly itself. The tab on a good used model runs well into six figures. This beauty was almost new.

Such has been the progress since I began, not only in flying machinery but in the ease of its operation. At least, this fellow's casual approach would suggest to an outsider that there's not much to flying anymore. Climb in, start up, and away you go just like in a car, but much faster and lots more fun. No need to check under the hood every time. Surely it is the ultimate toy.

On the first day my instructor showed me how to preflight an airplane. That is, closely inspect it outside and in. Of course, they didn't build them then like they do now. Those old Kinners were prone to sling oil and even rocker box covers now and then, old-fashioned wood props were known to split and you could see the patches where fabric had been torn by gravel thrown back on takeoff. The interplane wires became slack after hard landings; gas tanks and tires leaked, and so on and on. You had to be careful with those old airplanes.

He took me for an orientation ride, landed, and told me to meet him in 30 minutes. I was climbing back in when he stopped me. "Do another preflight,"

Braniff 720 tucks up gear for dash to Chicago.

he said. He meant the works, every nit-picking detail. "We know the tank was topped but make sure—and see if the cap's on tight. If not, let's find out right now and not when we're climbing over those trees." Made sense.

At the next school we flew the Harvard (AT-6/SNJ), a stout, all-metal brute with steel prop and big reliable Pratt & Whitney engine, and we flew it from paved runways. The preflight drill was the same. Overlook one minor item and you had to start all over. Though it was a whopping advancement over the primitive Fleet biplane, the Harvard likewise bore close scrutiny prior to flight. An oil-streaked belly or strong smell of gasoline could dictate a mechanic's investigation. Airplanes shouldn't drip or stink but sometimes they did back then.

The military throughout was hard-nosed about preflighting. You could bust a checkride without ever climbing aboard. There were numerous things to check outside; once they were committed to memory, transition went much easier. There was an equally detailed ritual to complete inside before reaching for the starter switches. Familiarity with a type did not make preflighting less essential. Taking off in a 172 with the pitot cover on, which I confess to, was an annoyance; the same oversight in a heavy B-24 at night (guilty again) was downright thrilling.

The airline was even fussier. We spent hours in the hangar quizzing one another on drain ports, strut extension, flap tracks, brake wear, and the plumbing mazes inside wheel wells. A fed could decide if you knew your business in five minutes, though one friend spent a miserable hour on the ramp before invited to exhibit his piloting skills. Don't be misled by a DC-10 flight engineer's seemingly casual walkaround which, incidentally, follows a mechanic's inspection. He knows what to look for.

The fact that a ship arrived without squawks was no guarantee it remained safe for departure. One gloomy old DC-3 captain was always annoyed at the company's question when we reported the field in sight, "Is your ship OK through?" "Well, it's OK right *now*," he'd respond, then tell me, "No telling what'll go wrong before we land." He had a point. The airplane that lost an engine on takeoff had performed normally on the inbound trip.

When the CAA (now FAA) required flight engineers on large airliners, I was among the first candidates to warm the third seat. Company rules required crews to report an hour before departure. The skipper could stroll in a few minutes late and the copilot right on the dot but the engineer with career ambitions had best be early. After learning if the proposed fuel load pleased his captain, he met the inbound two-striper to discuss any problems needing attention.

Then he inspected the airplane from nose to tail with no shortcuts because of rain or snow. The interior inspection was as painstaking. You wouldn't believe the number of little windows and doors hidden away in a DC-6 or Connie. Each was important. One mechanic regarded us novel newcomers sardonically, as though we distrusted his inspection. Eventually my chance came. Pointing to a prop I shook my head. There were no blade nicks, no leakage from the dome, no damaged deicer boots. "So what's wrong?" he asked. "It's in reverse pitch," I said. No big thing; had neither of us noticed, it would have corrected itself when the engine started, but it made the point. He was friendlier thereafter. When pilots and mechanics look at airplanes they see different things.

Ninety-nine of every 100 hours I've logged were in airplanes owned by other people. They paid me to fly and to fly their way. Their manuals dictated thorough preflight preparation including a personal inspection of the machine by me or, later, my copilot or engineer. There was an acceptance form to sign and woe betide the engineer who OKed an accumulator pressure in fact 100 psi low.

Our first personal airplane was a tired old Ranger-powered Fairchild 24 needing much TLC. I missed the army of supporting technicians that made airline flying such nice work. Now I was fuel boy, preflight mechanic, caterer, cleaner and operations clerk, all of whose duties required surprising effort. Preflighting took on new weight. An A&P checked me out, showing what to look for and how to spot things that mattered like the loose gas line not yet leaking enough to stain outside the cowl.

Our Bonanza was a big step forward, meaning there was much more to go on the fritz. More than once I was thankful for the strict upbringing that made careful preflighting a habit. Once, for example, we made a long trip requiring an early morning takeoff. Tanks were topped, the oil level checked, and all else was made ready the night before. At dawn we pulled the ship out and loaded the bags.

I did a quick walkaround including a final look at the tanks. The right was dry. Back in the dark hangar there was a broad stain on the concrete. Rare (but not unknown, I learned) factors—pressure, temperature, airplane attitude, whatever—caused fuel to siphon overboard. No doubt I'd have caught it on the gauges, but carelessness is infectious. It spreads from one area of attention to another.

My comparison of past with present is facetious. Little has changed. For all its advanced design and fail-safe features, the modern aircraft can fail its pilot, although it's usually the other way around. The fellow I watched knows better. His airplane had sat for four days yet, without a glance at fuel, oil, sumps, airframe condition or a check of flight controls (they never moved prior to takeoff), he flew. His cockpit check was likely as careless. He didn't earn his ticket with such an irresponsible display.

Flying is a privilege awarded those who demonstrate competence. The skills can be taught but the holder must supply the sense of responsibility their use demands.

28

Where is Bill?

THE PHONE RANG. "Len!" Though he rarely calls, I knew it was Charlie. For all the awful changes he endured, his drawl remains unmistakable. It is a link with a past that has become dreamlike, conjuring up memories of a wonderful and terrible time I rarely think about anymore.

A dozen of us Americans marched into the cold hangar on a winter day to receive, along with our Empire classmates, Royal Canadian Air Force wings. Three of us keep in touch almost half a century later.

Don came from New Jersey, a tall, lanky, reticent redhead with a keen sense of humor. He had logged a few hours in Aeroncas and wanted to fly fighters in combat. He would get his wish.

Charlie was from Alabama. Handsome, irrepressible and maddeningly cheerful, he was a wheeler-dealer. He could talk you out of your socks. Girls found him irresistible. "I bring out their mother instinct," he explained. The fact was, he wouldn't take no for an answer.

Bill was heavy-set, good natured and sported a moustache. He often spoke of his folks in Denver. He was vain but not insufferably so and played the role of Yank-in-the-RAF with fair success. For that matter, we all imagined ourselves to be adventurers, even soldiers of fortune, the only difference being that Bill *looked* the part.

We four remained together throughout training and for a year overseas. For reasons never explained, hundreds of us fledglings without operational training were posted to Egypt where there was no operational school. Utterly frustrated, we languished in a desert tent city. The U.S. Army arrived in mid-1942.

Bill, Charlie, and I transferred to the Army but Don, fearing an assignment to transports, maneuvered a return to England and fighter training. After the war, I learned the rest of his story. We three became C-47 copilots on the trans-African network just being established under contract by Pan Am.

While that flying was far less hazardous than what we had requested—P-40s—living conditions were miserable. Those days are remembered in terms of sandstorms, dysentery, sores that wouldn't heal, lousy grub, and grinding fatigue. When I ran into Bill or Charlie we'd wonder how Don was faring back in Blighty.

Twenty months after sailing from Canada we were rotated back home. Charlie and I waited for Bill who was due back from India. We had shipped out together; we would return together. A message came in: Bill had an accident. No survivors. Charlie and I left that evening on the long trip home.

Charlie went to a Dallas ferrying base. I drew the same lackluster duty at Memphis. Don's sister wrote saying he had been shot down. No details. After six months—the minimum rest period for overseas returnees—Charlie was sent to the China-Burma- India theater to fly the infamous Hump. I was sent to a C-46 school whose graduates also went to the "Rock Pile."

At that early stage of the war our ferrying groups were full of pilots itching to go overseas, or so they claimed. The rush to send us early returnees back was due to our unpopularity. Our campaign ribbons (which only said we'd been there) and the two pairs of wings made the stateside heroes uncomfortable. My CO was a local type who had managed a home time assignment. We didn't hit it off.

I was kept in Nevada to instruct. Charlie was not as fortunate. He crashed in a C-109, a flying bomb created by converting a B-24 to tanker aviation gas. Its crews wore tennis shoes to avoid sparks, yet their ships exploded on the ground and in flight for no apparent reasons. Charlie crashed on takeoff in India. He ran from the wreckage hideously burned, the sole survivor. Doctors gave him scant chance of living. They would come to know of his indomitable spirit. "I got scorched," he joked when I first entered his room, then explained in detail what had happened.

After four years in hospitals and literally scores of operations, he emerged, a hairless apparition minus ears and eyelids, blind on one side, his nose constructed with flesh from his arm. What remained of his fingers were frozen into claws. "I was lucky," he said. He invested his mustering out pay in rent property, mortgaged it to buy more, and so began a success story perhaps unmatched in American business, all things considered. Yet he often said he'd trade his amazing success for my airline job.

After a year on his own he took me up in his Globe Swift, the first of many airplanes he would own. His recent call was to ask for advice on business twins, advice I could not offer, but I referred him to someone who could. He later sent a note describing what he bought. The scrawled signature, "Chas.," is exactly as he signed his enlistment papers. His letter included the usual invitation to visit at any of his several homes. Charlie made it big and earned every million of it the hard way. And I haven't told you the half about him then or now.

After the war I met Don in New York where he worked on Wall Street at a job he detested. Most of us were finding in the aftermath that civie street offered nothing to match what we had known during the previous five years. Don said little about his war experiences, but after my repeated requests he typed out a brief summary which, not surprisingly, he entitled, "The Story of a Non-Hero."

Upon returning to England he requalified in the Harvard then checked out in the Mustang I with an Allison engine. After completing operational training he was posted to No. 268 Squadron at Tangmere. The assignment was to intercept hit-and-run raiders plaguing towns on the Channel coast. When the outfit was relieved and sent to Scotland for a rest, Don transferred to the 357th Fighter Group, 8th Air Force, also P-51-equipped, and was engaged in high level bomber escort. "I sometimes regretted that," he said. "I never again felt the camaraderie we knew in the RAF." I knew exactly what he meant.

He named his Mustang, "Blythe Spirit." He's vague about details, but noting the targets mentioned in passing Regensberg, Augsburg, Frankfurt, Hamburg, Berlin—I could read between the lines. After 60-odd missions, while on a sweep near Paris, a number of 109s were attacked. Don was firing at one when another put a cannon shell in his engine. The cockpit filled with smoke. He jettisoned the canopy and bellied into a wheat field. After walking and hiding for 10 days he met advancing American troops and was flown back to England.

Don's retired in New England after 30 years with the FAA, working in airway traffic centers and on research projects. He and his English war bride spent a night with us. We didn't begin to catch up.

And Bill. After the war I went to see his parents in their small home. He would have done it for me. I slept in the room that was his as a boy attending East Denver High. A gold star hung in the window. He was their only son. Answering their many questions, I recalled our training days, the long voyages to Liverpool and Suez, our months in Africa, making them sound more fun than they were. I said Bill had been completely happy and doing exactly what he wanted right to the end. That part was completely true.

His mother handed me a letter from Washington: Did the family wish the remains brought home? "What would you do?" she asked. I said he would have wanted to remain among his fallen friends. Need this heartbroken woman gaze upon a flag-draped casket? I believe Bill would have thanked me.

Over the years we lost touch. I wondered what had been decided. I wrote to the American Battle Monuments Commission, giving Bill's name, rank and serial number. Where is his grave? No record of him could be found, nor were the other agencies I was referred to able to help. No mention at all in that long list of names?

That bothered me. It still does. Bill was a great fellow.

29

The early airline days

ONCE THE WRIGHT BROTHERS HAD SOLVED THE RIDDLE of controlled flight, it was inevitable that their invention would be used both as a weapon and a vehicle of commerce. It filled both roles within a decade, a hint of the rapid progress of the new science would see.

The military had long before found the balloon useful for observation. Intelligence gathered by army aeronauts contributed to the French victory in the Battle of Fleurus in 1794, one instance of such employment. Victory belongs to the general who can guess what's happening on the other side of the hill, said the Duke of Wellington a century earlier. Aerial observers looked beyond the hill.

The commercial possibilities of flight also were obvious from the first. Aeronauts inflated their linen bags for exhibition, passenger hops, and scientific exploration of the mysterious atmosphere. Happily, people would pay to soar aloft or simply to watch. But, a slave to the fickle winds, the balloon's potential as a weapon or for practical transport was hopelessly limited. It was a fascinating novelty, no more.

The Wrights revealed the manner of future flight. Again, the military was interested and so were the profit-minded. After chugging along in a primitive aeroplane over hill and dale, going directly from here to there, it took little imagination to see the possibilities of air travel. Who would travel by road or rail if he could go like this? Grandiose plans were made and stock issued in air transport companies but the visionaries had no conception of the problems involved. All of it would come to pass, but only after a costly struggle lasting nearly half a century.

On New Year's Day, 1914, a small open cockpit flying boat left St. Petersburg for Tampa to establish the world's first airline service. Daring its 90-day life, the line carried 1,200 passengers without mishap and completed 91 percent of its schedules. A modest effort, the St. Petersburg—Tampa Airboat Line accomplished what few of its successors could during the next quarter

Retired DC-7C sits in weeds awaiting sale. Convair 440 in background.

century—it earned profits. In passing, it bears noting that the airplane's role as an airliner predated its first hostile use by six months.

Despite the historic first, the United States lagged behind Europe in post-World War I air travel development, concentrating instead on its airmail services. The first daily sustained passenger service began between London and

Paris on August 25, 1919. Within three years, five lines competed on the prestigious route. The first sustained daily passenger service on this side began on May 23, 1926, when a Western Air Express Douglas M-2 left Los Angeles for Salt Lake City with fares aboard. Though the question as to which of our airlines is the "oldest" will never be resolved to everyone's pleasure, Delta, through its merger with WAE's successor, Western Airlines, now flies the route with the longest record of uninterrupted service.

Scores of operators entered the picture with a variety of aircraft, mostly single-engine monoplanes. Their services were sporadic, noisy, uncomfortable, risky, and nearly always doomed to failure from the start. There was little about them to lure passengers away from the security and comforts of Pullman travel. To be sure, flying was novel. Being able to drop, "I flew back from Chicago this morning," into party chatter had the same electric effect in the late Twenties as, "We returned on *Concorde*," would have in the late Seventies.

But the odds against economic success were overwhelming. There were few adequate airports and virtually no navigational aids. Operations ceased at sundown. The equipment available was unsuitable and manufacturers were reluctant to risk development money on the larger and more efficient types needed for airline service.

Most discouraging perhaps was the public's fear of flying. Lurid press accounts of accidents reinforced the popular conception of flying as downright suicidal. The curious flocked to airports to watch but few bought tickets. They admired the uniformed pilots (who joked that starvation was their greatest risk) and the brave souls who with feigned nonchalance climbed aboard behind them.

Inevitably, there followed a series of failures, mergers, and acquisitions. The resulting confusion rivaled anything seen since the recent airline deregulation. In fewer than four years, for example, 12 small carriers through mergers became American Airways, later American Airlines. A respected historian's intensive research produced no clear picture of the mad scramble for survival during those turbulent times. His offer to pen a definitive history of one now-major carrier was refused when he asked to examine early business records. Another of today's giants likewise declined when similarly approached. There are skeletons aplenty in airline closets.

In May, 1927, a quiet, unassuming airmail pilot flew nonstop from New York to Paris, giving the struggling young industry an unexpected and welcome boost. No other individual achievement in American history before or since has resulted in the worldwide acclaim accorded Charles Lindbergh for his astounding feat in the *Spirit of St. Louis*. He then flew the *Spirit* to all 48 states, landing at 78 cities. Before the year ended, the "Lone Eagle" flew nonstop from Washington to Mexico City and on to 16 Latin American countries.

Overnight, people became, if not air-minded, at least more air conscious. It was suddenly clear that a well-designed airplane could cross great distances quickly and dependably, when expertly handled. The effect was profound though not immediate. The surge was not to buy tickets but airline stocks.

Banks, automobile builders, railroads, and private investors wanted pieces of the action. Particularly pleased with the sudden demand for its stock was the Seaboard Air Line, a railroad with no interest in aviation.

A consortium including the Pennsylvania Railroad formed Transcontinental Air Transport a year after the Lindbergh flight and made him its technical adviser. The endorsement was beyond price—TAT would be "The Lindbergh Line"—but the value of the hero's name was more than matched by the advice he gave regarding routes, facilities, pilots, and equipment. The decision to buy three-engine Fords and not fly at night reflected Lindbergh's conservative approach after considering all factors. He envisioned as safe and dependable an operation as the state of the art made possible, and he acknowledged its limitations.

An air-rail coast-to-coast service was inaugurated on July 7, 1929, amid much fanfare. Lindbergh pressed a button in the governor's office in Los Angeles, a light flashed in New York's Pennsylvania Station, and the *Airway Limited* rolled away into the night while a band played, "California, Here I Come." The train reached a specially built terminal near Columbus, Ohio, soon after dawn where passengers walked to two Fords and were airborne in 20 minutes. After four stops, the airliners reached Waynoka, Oklahoma, at dusk. There passengers transferred to the Atchison, Topeka & Santa Fe Railroad for the overnight run to Clovis, New Mexico, where a pair of Fords completed the trip to Glendale with three en route stops.

The next day Lindbergh commanded a Ford out of Glendale to inaugurate the return schedule. A crowd of 100,000 watched the *City of Los Angeles* take off and turn toward the rising sun. Great pains were taken to keep the airliners moving with railroad precision, exactly 15 minutes being allowed at each stop for fuel and to let deafened riders stretch their legs. The New York—Los Angeles journey was timetabled at about 48 hours, 24 hours less than the fastest train, with fares averaging $370 one way.

To publicize the new service a TAT Ford christened by Amelia Earhart was exhibited in the main concourse of Penn Station. Thousands climbed the platform alongside the gleaming Tri-Motor to peer in at the wicker seats and cluttered cockpit of the *City of Columbus*, among them one seven-year-old boy who thought it the most marvelous thing he had ever seen.

The timing could not have been worse. Four months later came the stock market crash and $100 a month became a living wage. After 16 months the air-rail experiment was discontinued, having never earned a profit. Kept alive by airmail subsidies, TAT struggled on, surviving to become one of the "Big Four"—TWA. Change the names and TAT's early history is the record of every airline during those lean times.

Ford, Fokker, and Stinson trimotors were replaced by the roomy but lumbering Curtiss *Condor* with which American Airlines offered sleeper service. Lockheed's 10 *Electra* also offered twin-engine reliability and was faster but only carried 10 passengers. Boeing introduced its Model 247, an advanced design with retractable gear able to outrun military fighters of the day. A marked improvement over its competitors, the 247 was very nearly the airplane needed to earn profits carrying passengers.

TWA asked Douglas in 1932 to design a metal trimotor capable of 146 mph with 12 passengers. The builder countered with a twin-engine design, the DC-1, which was accepted with Lindbergh's proviso that it be able to take off with a full load from any of the line's airfields. The sole example built was an immediate success. Orders poured in for the stretched (to 14 seats) DC-2.

American Airlines asked for a still larger model with 21 seats but Douglas, fearful of spoiling the DC-2's success, was reluctant. A firm order for 20 did the trick. The first DST (Douglas Sleeper Transport) service began in September, 1936, when AAL's *American Mercury* departed New York. Thus began the incredible story of the DC-3, as most later models were designated.

Although the Douglas marvel fell short of American's claim of a 190-mph cruising speed and 2,000 mile range, its cabin was indeed "the most spacious, the most luxurious . . . the quietest." More importantly, the DC-3 marked the beginning of the end of unprofitable passenger operations. When Pearl Harbor was attacked five years later, 80 percent of the airliners in domestic service were DC-3s. Of the "grandfather" lines, National alone did not order any DC-3s.

The first chapter of the airline saga ended with World War II after which hundreds of efficient aircraft became available for air transportation and there were thousands of well-trained pilots eager to fly them. While much hard work yet lay ahead, there was no return to hungry days of the Twenties and Thirties.

Regrettably, the industry has shown little interest in its beginnings. Some major airlines today cannot list all the types of aircraft they have operated or when routes were acquired or dropped. The editor of one line's house organ was told to write a brief company history. The archives supplied were a shoe box of snapshots and old timetables.

Years ago an industry museum was proposed. At the time the airlines well could have afforded to subsidize such a worthwhile record of American achievement by donating aircraft, records, photographs, and supporting the collection's maintenance. Imagine walking down the line to examine a Vega, Ryan B-5, DO-2, Travel Air 6000, 247, Electra, Stinson SM-1, and other types with which it all began, not to mention the great aircraft that served since their day. But nothing came of it and now it's too late. That's a shame, isn't it?

30

Full circle

WHEN I BECAME SERIOUSLY INTERESTED IN FLYING, most passengers rode the Boeing 247D, Lockheed 10A Electra, and Douglas DC-2 airliners with two engines and two pilots. All three were marked improvements over the trimotored Fords, Fokkers, and Boeings they replaced, yet somehow they seemed a step backwards. Were not three engines better than two? I marveled at the big Sikorsky S-40s and S-42s at Pan Am's Dinner Key base at Miami. They had four engines as did the sleek Boeing 314 Clippers and Short Empire flying boats I saw anchored at LaGuardia Field during the 1940 World's Fair. Surely that was the way to go.

The Germans seemed to me to have the right idea in 1929 with their three-deck Do-X flying boat, a behemoth with 12 engines that once rose from Lake Constance with 169 souls on board. And there was the Russian eight-engine Maxim Gorky, an unbelievable monster with 65 feet more wingspan than a 747. In retrospect, both were freaks but they pointed the way, or so it seemed to me. Germany's *Hindenburg* lumbered back and forth across the Atlantic; airships most certainly would figure in the coming air age, probably as freighters.

Engines were amazingly reliable by the late 1930s and there were pictures of DC-2s flying over Manhattan with one prop stationary. Two might do for 14 passengers but they'd need six, eight, or more on airliners of the future. Sure enough, along came the Douglas DC-4, Lockheed Constellation, and Boeing Stratoliner, each with four engines. The Four and Connie sat level, a return to the gear arrangement used by many pioneers. The 1909 *Jane's* pictured many airplanes sitting level on three or more wheels.

The war halted further transport development; existing designs were mass produced for military use. When it was over, the seemingly inevitable trend toward bigger-with-more-engines resumed. Hughes flew his monstrous wooden boat and the British their Brabazon, each with eight engines. The enormous Saunders Roe flying boat had 10 engines driving eight props while

DC-8 engine cover reflects mechanics' humor.

the Air Force ordered the B-36 with six—and later, 10 engines. A 400-seat transport version of it was flown. No crystal ball was needed to predict the future.

A semiserious joke of the time went, "I want to fly an airplane on which, when the flight engineer says, 'Captain, we've got an oil leak on Number Six,' I ask which side?"

The jet engine changed everything. The propeller followed the tail wheel into oblivion. The Comet, 707, and DC-8 appeared with four engines, not six or eight. In fact, the turbine wizards' success at packing more and more thrust into the same slim cowlings brought about a reversal in trend. Surprisingly, the 727 flew with three engines and the 737 and DC-9 were twins. While the last two are considered "small" airliners today, stretched models of them seat as many fares as the first 707 and DC-8. Who would have believed it? And who would have predicted the 757, able to carry 239 coach passengers, and the 767 with up to 290 seats, both of them two-pilot, two-engine aircraft, much less that the latter would be certified for transatlantic schedules?

The evolution of crews likewise upset all predictions. The complexity of postwar domestic airliners, increasing traffic, and a series of accidents led the CAA to require a flight engineer aboard larger equipment shortly after the war. An international crew traditionally included a navigator, engineer, radio operator, and often relief crewmen to spell them and pilots on long distance runs. The radioman went first, then the navigator, and finally the flight engineer. Two pilots now fly 550-seat jumbos, with relief pilots carried on extremely long runs.

The evolution of turbine power produced equal surprises. The original turbojet's rapid spin was geared down into a shaft and the prop was back in style. They said the resulting turboprop was an interim engine to be phased out when pure jets took over. Then came the more efficient and quieter turbofan, which pulled as much as it pushed. In effect, its fan was a series of multibladed fixed-pitch propellers. And now here comes the propfan with curious scimitar-like counter-rotating props at the rear of a turbine nacelle.

The endless years I spent on the wrong side of the airport fence are often referred to as the "Golden Age" by buffs too young to remember those lean times. It was only golden for those with gold. Private flying was a sport for the wealthy. A few large corporations had airplanes—then as now unmarked so as not to raise the eyebrows of stockholders who witnessed their grand arrivals—but those mouth-watering Stinsons and Wacos were far beyond the means of the average Joe.

The war gave civil aviation a tremendous boost. What theretofore had been an adventure for a daring few suddenly seemed to make sense. Millions of young men had flown—or flown in—airplanes by the time it was over. The postwar transportation revolution did not alter the picture as rapidly as most prophets promised, but the changes it brought have been more profound than they expected.

Events did not bear out two popular predictions. First, that private flying would became almost as common as driving. There would be "a plane in every garage." Magazine ads depicted Dad walking out to a neat four-seater waiting on the back lawn while Mom and the kids waved good-bye. It was Monday morning and he was off to the office. Just where he would land in the city was not made clear.

Or, there was the air-age family picnicking beside a small lake with their flying sedan in the background under the trees. Every small town would have an airstrip and most would have shuttle flights to the nearest big airline field. Few gave these insane notions five minutes of serious thought, in fact, many the pilot veteran who invested his wartime savings in private aircraft and engine builders' stocks.

And, some supposed military experts predicted an early demise for manned fighters and bombers. Within a decade or two all offense and defense would be accomplished with guided missiles. Except for limited transport and administrative flying, air forces of the future would consist of wingless technicians sitting underground with their fingers near ominous red buttons.

Still another unrealistic premise was that anyone with a surplus military transport or two could make money hauling passengers. The non-sked era ensued. With C-47s, C-46s, and C-54s bought for a dime on the dollar, new "airlines" were formed by entrepreneurs whose enthusiasm was rarely matched by adequate capital and business acumen. Their discounted fares could not sustain long-term growth and profits. The resulting competition between them inevitably led to corner-cutting; after a string of bad accidents, the feds clamped down and most of the upstarts folded.

Developments no one predicted have often surprised we hopefuls who lined the fence before the war. The most optimistic among us didn't envision the 747 or F-16 or a field like O'Hare or the scope of general aviation in the 1980s or that any of us would someday own personal airplanes. Most of our youthful predictions were proved wrong, not that the record of the experts has been much better; there have been many surprises, not all of them good.

Everything now is so different from what it used to be, and yet . . . there are serious plans afoot to build a 200-passenger airship, manned fighters and bombers are being ordered by the hundred, and the prices of personal aircraft have soared to the point almost none are being sold. Flashy new "airlines" flying surplus airplanes lure riders with ridiculously low fares and collapse before we learn their names. Senseless competition has once again produced corner-cutting, resulting in huge federal fines.

And, if you buy a domestic ticket, the chances are you will ride a twin-engine airliner flown by two pilots. In a few years it might have two propellers.

I believe this is where I came in.

31

Escape to Switzerland

MY MINOR ROLE IN THIS STORY BEGAN when someone pointed to an AT-6 overhead—minus its rudder. The ailerons apparently provided a modicum of directional control because the unlucky pilot made a shallow turn onto long final and dropped the gear. We held our breath. He nearly made it, but as he crossed the fence the nose yawed wildly, a wing dropped, and the airplane cartwheeled. A qualified observer would have been hard pressed to identify the type from the wreckage, yet the shaken pilot walked away without a scratch.

It occurred that a book with pictures of similar crashes might have appeal. I asked friends for help and ran a small ad in two magazines. The response was surprising; soon I had 300 crash photos, some almost unbelievable. The better half was published under the title, *Crackup!* The book did fairly well—about 16,000 copies were sold—and eventually went out of print. I rather expected better sales. Apparently many collectors were repelled by the whole idea of crashes even though every accident pictured was nonfatal. The preface made it clear that none resulting in death was included. The intent was to show that major air accidents are often survivable.

That was more than 20 years ago. This story actually began 25 years earlier, on April 24, 1944, at an airfield in East Suffolk, England. Station 119 was the home of the 95th Bomb Group, which dropped nearly 20,000 tons of bombs during 314 combat missions. It was dicey duty. By the war's end, 157 of the 95th's aircraft were missing in action, with 39 others posted as "other operational losses."

It was still dark when 2nd. Lt. Max Wilson and his crew were roused for breakfast and briefing. It would be their fourth mission. The target: a ball bearing works at Friedrichshafen in southern Germany, across Lake Constance from Switzerland. Field Order 315 directed 754 bombers to raid the target, accompanied by 867 fighters. The 95th contributed 23 aircraft to the effort. Ten would sustain battle damage and two would not return, one of them was a B-17G, serial number 42-31632, commanded by Lt. Wilson. Among his 10-man crew was 19-year-old Staff Sergeant Ron Grove of Waverly, Missouri, the tail gunner.

Things went badly from the start. Number 4 engine overheated and was shut down. Wilson increased power and maintained close formation. Approaching the target, the group was attacked by Focke-Wulf Fw 190s, one of which put a cannon shell into the Number 2 engine. The bomb load was dropped on the target and the battered Fortress dropped out of the formation, unable to maintain speed. It was then discovered that Number 1 engine had been hit and was losing oil.

Unable to maintain altitude on one engine, Wilson headed for Switzerland and ordered his crew to bail out. Two parachutes were found to be flak damaged—all the crewmen elected to remain aboard while the pilots crash landed. An open field was sighted near Dubendorf. Believing the ground to be hard frozen, Wilson extended wheels and flaps.

On touchdown the left gear collapsed, having been damaged by the 190's cannon fire. The Fort slewed to the left, slid into a radio substation, and came to a stop against its steel tower. Fuel spilled from ruptured tanks but there was no fire. The bruised crew climbed out to be interned for the duration.

During the war about 250 Allied and Axis aircraft sought refuge in neutral Switzerland after battle damage made return to their bases improbable. (Wilson's B-17 was one of 14 to end up in Switzerland that day.) More than 300, including 68 B-17s and 59 B-24s, landed in Sweden. There were dark hints that not all had been shot to pieces as had Lt. Wilson's Fort, rumors that more than one landed though still airworthy and with fuel in its tanks. Whether or not this sensitive matter warrants reviving 45 years later is debatable. In any event, it became the subject of the television documentary "Whispers in the Air."

Internees enjoyed considerable freedom but were accompanied by guards wherever they went. Seven months later Ron Grove and a fellow internee from the 384th Bomb Group escaped to France and were flown back to their units in England. No further combat flying in the European Theater of Operations (ETO) was permitted as an escapee might be shot down and recaptured. His contacts and escape route had to be kept confidential. Grove was returned to the States and was sworn to reveal no details of his getaway until the war ended, even to his family.

And that might have been the end of the story had not his son, Gerald, become intrigued by World War II history and his dad's unusual experiences. He pored over his father's records and pictures and wondered what had happened to Lt. Wilson and the rest of the crew of B-17 No. 1632. Ron Grove had not heard from any of them since his escape in 1944. Had the other nine young men survived the war? If so, where were they now? The more his son read about the ETO air war, the more he wanted to know.

Gerald spotted *Crackup!* in a hobby shop in 1984. One chapter entitled, "Escape Into Switzerland," included 25 pictures of battle damaged B-17s that found refuge in that neutral country, among them B-17G No. 1632 of the 95th Bomb Group. "Few moments in my life could match that one for excitement," says Gerald. He wrote asking if I could supply a copy of the picture.

The Swiss photos had been loaned by Steve Birdsall, an Australian historian who authored a book on the B-17. I forwarded the request to Steve who sent the original photograph to Gerald. That did it. He was determined to find the rest of the crew. "Reading history and building models taught me patience and persistence," he said. "I would need both."

After a considerable search, he discovered the right waist gunner's address and phoned, only to learn he had died years earlier. His widow was able to supply three phone numbers, however, including that of Lt. Wilson, the pilot in command. Further research has accounted for eight of the 10-man crew. The search for the remaining two continues.

Ron and Gerald Grove flew to London in 1989 and drove to Horham, the tiny village adjacent to Station 119. The site was long since returned to farming. The outline of runways and perimeter tracks can be seen from the air but little evidence of its wartime use can be found while walking across the open fields.

When the headquarters flag was lowered for the last time in 1945, it was given to the local church where it is displayed today. In the village stands a gray marble monument the shape of a B-17 rudder with the 95th Group's "Square B," a silent reminder of a war most people cannot remember. On the inscription is a runway diagram of the airfield that is now but a memory.

At Duxford, an airfield built in World War I and home of the USAAF 78th Fighter Group in World War II, the pair were given a close look at B-17s being used in the filming of *The Memphis Belle*. Duxford's 75-year-old wooden hangars house the Imperial War Museum's magnificent collection of military aircraft, many of them American and British types familiar to every American airman who was in Britain during the early 1940s.

The Groves then flew to Zurich, rented a car and drove to nearby Dubendorf, wondering if they could locate the crash site. When they stopped at the Swiss Air Force Museum to seek assistance, Gerald looked around, spotted a crash picture, and realized they had already found it. The open field where Ron Grove's crippled Fortress slid to a stop 45 years earlier is now a Swiss Air Force base.

Gerald has accounted for eight of the 10 crewmen, three of whom are deceased. He is still looking for James W. Pletcher Jr., the ball turret gunner from Wilkinsburg, Pennsylvania, and David K. Oates, left waist gunner from Seattle. If a reader can provide any clue about either, I'll pass it along to him.

Historical research is dull, plodding work that is occasionally rewarded by surprises. Gerald was watching when "Whispers in the Air" was televised. The controversial documentary began with a film clip of a badly shot-up B-17 crash landing in Switzerland. The left gear collapsed, the bomber slid along and crashed into a radio tower. It was B-17G No. 1632 of the 95th Bomb Group.

32

The recip days

THE MECHANICALLY-MINDED BOYS OF MY DAY fell head over heels for the internal combustion engine. The rhythm of intake, compression, combustion, and exhaust strokes was easily visualized, with valves lifting and falling and plugs firing at precisely the right times. It was alive; it spoke, telling the knowledgeable what was right or wrong. When mixture and spark were right, its bark was sweet music; it spit flame or smoke or lagged when they weren't. In fact, the glorious sounds of short-stacked aircraft engines made flying all the more appealing to us.

"Listen to 'em," coached a patriarch who had flown the mail in DH-4s. "They'll tell you how they're doing." Picking an occasional discordancy from engine roar or noticing a minor throb in the rudders was a knack slowly acquired. You learned by paying close attention and followed through by checking gauges, looking long at each engine and easing back on the carb heaters; roughness often traveled along cables to your fingers. You chatted with a passenger while studying the exhaust flame outside his window.

Jet thrust is managed with a single lever. There is one adjustment for climb and another for cruise—period. The operation of piston engines was quite different with throttles, prop controls, mixtures, mags, cowl flaps, oil coolers, and carb heat to check before takeoff. Throttle and mixture adjustments were frequently made in flight.

Icing problems were solved by book procedures and tricks learned the hard way, not with a switch marked, "Engine Heat." An icing dilemma was always more easily prevented than solved. The tender loving care that kept big recips smoothly running began when the gear doors shut. The rate of climb stipulated for noise abatement might send head temps soaring in July. It was, "Let 'em cook" or explain to the FAA—you couldn't win. Most of us chose to explain.

To conserve fuel, the jet pilot remains at high level to the last minute then rapidly descends with thrust levers at idle. That was extremely bad form with

Electra props made obsolete by turbojet powerplants.

pistons. Rapid cooling warped cylinders and fouled plugs. Let down through cold cloud in idle and you might collect more ice than cooled engines could melt. A DC-7 rule of thumb was to maintain manifold pressures at least equal to rpms. That is, at 2,100 rpm, pull no less than 21 inches—and advance throttles now and then to confirm that all four Wrights remained wide awake. Climb and descent required careful planning and continuous monitoring of engine gauges.

Heavy rain soaked ignition harnesses causing misfiring. You waited for air flow to dry them—once you got back into the clear. Failure to wet down a DC-3's prop blades with alcohol produced severe vibration and an unnerving gunfire of ice slivers slung against the fuselage.

Historians refer to the -3 as the first airliner to earn profits for its owners. Powered by two Pratt & Whitney 14-cylinder, or Wright nine-cylinder, radials, it didn't quite live up to its three-miles-a-minute billing; however, if you divided mileage by three, you'd usually have the destination in sight when the time ran out. Incredibly rugged and dependable, the Three has outlasted many of its successors. More of them are flying airline schedules today than Convairs, DC-4s, -6s, -7s, -8s, or 707s. Who would have believed it? Those of us who flew DC-3s, that's who.

The old girl was admired by pilots and passengers alike. A friend logged 12,000 hours in DC-3s without an engine failure. Even when the military added 2,500 pounds to its civil takeoff weight of 25,000, it could usually make it around the pattern when one quit at 100 feet, if handled with a watchmaker's touch. Twenty-one fares rode in reasonable comfort, lulled by hypnotic engine drone and vibration, down low where there was something to see.

After lunch, a form with the blanks filled in was sent back: "Welcome aboard Flight *22*. We are cruising at *6,000* feet. Our groundspeed is *172* miles an hour. We expect to arrive at *Wichita* at *8:47 a.m.* Please tell your hostess, Miss *Engle*, if there is anything we can do to make your trip more pleasant.—Captain *Art Gibson*." All in all, and weather cooperating, it was a most pleasant way to travel.

Constellations and DC-4s serviced premium runs after the war. The Connie was pressurized and faster, yet the converted C-54s (which most DC-4s were) served faithfully for years and until Douglas launched its DC-6. Like the DC-3, the -4 was popular with crews and fares. It waddled across the airport to a chorus of underfloor hydraulic groans and squeals, provided a pleasant enough ride by standards of the day, and settled ever so lightly onto its long struts.

While braking on short slick runways inspired a technique not in the book—rocking wings to dig in the mains—you'll find few veterans with bad words for the faithful Four. One of ours, flying in the clear between anvil heads, ran into intense hail. The nose was caved in, windshields shattered and deicing boots torn from wings and tail, yet the crew reported it handled normally throughout approach and landing. Rebuilt, it was returned to service.

The DC-6 looked like a stretched -4 but was in many ways a new design. It was pressurized to cruise as high as 25,000 feet, had two-stage blowers, water injection, automatically synchronized and reversible pitch props, heated wings, and heated tail. It also contained lethal flaws that, after several accidents, caused it to be grounded for modifications. Thereafter it performed with distinction on domestic and international routes.

With DC-6s and stretched Connies, airlines offered faster schedules and improved cabin services including sleeper accommodations (about which there are interesting stories to be told). Neither type could guarantee the above-the-weather service too often advertised. Their 20,000- to 25,000-foot cruising altitudes avoided the summer chop familiar to DC-3 riders, but put them waist high to Kansas thunderheads. Weather radar was installed and again ad writers pulled out all the stops to the annoyance of pilots. "If you've got radar, why the rough ride?" was a tough passenger question to answer.

Convair's 240/340/440 series was intended to serve the short-to-medium distance market as were the Martin 202 and 404. We got Convairs and they were remarkably successful. With more than twice the DC-3's seats, it moved along at a respectable clip and was fun to fly. It was powered by the same Pratt & Whitney R-2800 radials as the DC-6, in fact, high time Convair engines were rehung on our -6s to run out times between overhaul. If a Convair engine packed up during initial climb, its prop automatically feathered. You had three seconds to decide if the engine had actually lost power or if the feathering system had malfunctioned, in which case you grabbed the red button.

Integral loading steps folded out just ahead of the left prop. This made us think. They were lowered by the hostess but it was the copilot's responsibility to raise them and ensure the door pins were fully extended. Among other Convair novelties was a P.A. system. "I'm a pilot, not a tour guide," grumbled the typical captain, handing the mike to his junior. The Convair afforded its riders, particularly those in back, a decidedly noisy ride.

The colorful era of piston-powered American airliners reached its zenith with the graceful 1649 Constellation and handsome DC-7C, the first copy of which we received in 1956. A lengthened center section moved the inboards five feet farther from the fuselage. This provided room for more fuel, reduced cabin noise—and made it a handful when a check pilot chopped an outboard on takeoff. But its near-perfect control feel made it a delight to fly. The -7C almost found its own way down the ILS and was easy to land if you maintained a modicum of power into the flare. Chop the throttles over the fence and you'd arrive like a load of sand.

The exhaust gases from its huge Wright 3350s were ducted to three power recovery turbines connected through fluid drives to the crankshaft. These PRTs became white hot during climb and, if memory serves, spun at 19,000 rpm. The glow from Number Three in climb was enough for the copilot to do paperwork by without a flashlight. The Pratt & Whitney 4360, an incredibly complex four-row 28-cylinder monster developing 3,500 hp, powered Boeing's awesome Stratocruiser. That blunt-nosed monster featured a small lounge beneath the main cabin, prompting one competitor with Connies to advertise, "No stairs to climb!"

Flight engineers of these advanced types monitored powerplant health with electronic analyzers with which the firing of individual plugs could be studied. A "cold jug," for example, (a cylinder in which neither plug was firing) usually dictated a precautionary shutdown for it might contain a valve lip hammering a hole in its piston. Such a breakdown did not always produce vibration or decay of instrument readings noticeable to a crew busy with other matters. In fact, major internal damage requiring complete overhaul could be suffered by an unmonitored engine before anything amiss was detected. The old DC-3 clues did not always serve.

Considering the rigorous demands of airline work, piston powerplants performed extremely well in every corner of the globe and in most weather conditions. They required considerably more attention than today's turbines and, of course, there was more time to give them. We moved much slower and shared the air with a fraction of today's traffic. Except at fields such as Chicago Midway during rush hours, we flew straight to a field, dropped the gear, and landed. When a different en route altitude was desired, it was usually had for the asking. Holding was a procedure practiced in the Link trainer but rarely needed in the real world. We often flew an entire summer month without filing IFR.

And how does the reciprocating era compare with today? That is, from the standpoint of flying a schedule from here to there, and ignoring (if that's possible) the current instability of cockpit employment. Today's flying is far more regimented and precise than during the heyday of piston aircraft. The training is more intensive, the work in many ways is more demanding and less forgiving. The emphasis continues to shift from captains to computers until something goes wrong, at least. And, statistically, it is a considerably safer career than it used to be.

Listening to today's pro, I wonder if he derives the personal rewards we knew in the recip days. There was real satisfaction in successfully coaxing four big moody, often irritable, piston engines through winter conditions, shooting raw data approaches to 200 feet and wrestling too many tons to a walk on short, slick runways. We were proud to be part of that and it was great fun to boot. Is it still that way out on the line?

33

No cast-iron guarantees

IF YOU WERE SEATED next to Bill Loverseed on a flight to London, you would soon guess he was on his way home. It's not the accent that places him; it's more such British idiom as, "quite," where we say, "exactly," and "aircraft" and "aeroplane" instead of airplane. He'd probably not tell you what he does unless you asked.

Different circumstances brought us together. An English broker called, asking if we'd like to sell our airplane. Perhaps, we said. Then please send a description—engine and airframe times, list of equipment, status of logs, etc., and some color snapshots, including one of the interior of the rear fuselage shot from the baggage compartment. Apparently that would verify that it was free of corrosion.

During the next call we discussed money and the paperwork involved in transferring title. It was unusual in this case. "Our pilot will pick it up in a few days," said the broker.

"Tell me about him," I said, more than a little curious.

"He's ex-RAF, actually; splendid fellow. You'll like him."

Bill looks 40, admits to 55, and is as advertised. Soft-spoken, self-effacing, and deliberate, he is immediately likable. We met at our field and after his painstaking inspection of the airplane and logs we went flying.

"You've got time in one of these?" I asked.

"Yes, a few hours about two years ago," he said. I've flown with many pilots in a fair range of types. A few were outstanding, most were solid FAR-abiding citizens and another few are best forgotten. There is something about a pilot at work that quickly gives him away. It's a combination of clues: composure, attention to detail, technique, priorities, mannerisms, and what he says.

He climbed to 2,000 feet, stalled clean and dirty, checked radios, instrument readings, fuel flow from each side, and asked about carb heat. "I use it every time, summer or winter," I said.

He nodded. "Yes, I rather think I will." I guessed that he was already decided on that and was curious about me. Knowing something of a seller's habits can reveal things about his airplane that don't show on the Hobbs meter. We talked about fuel consumption and mixture leaning in detail, then landed. He flew with that consideration of the machinery that marks the pro. He put it down, fully stalled, with the feather-light touch I get one time in 20.

"It's a nice little aircraft," he said. "I'll leave at first light." That would be the beginning of a journey that would see our Cessna 172 delivered to a new owner in England. I'm interested in a pilot whose work includes flying a small single-engine airplane across the North Atlantic in midwinter.

With tape recorder running I drew him out. Loverseed was born in Egypt, the son of an RAF pilot. Orphaned as an infant, he lived with relatives in London. When the war began, children were evacuated. "They sent me to a village on the Channel in West Sussex; that turned out to be ridiculous for it was bombed more than London. I was eight and had a marvelous time, actually. The Battle of Britain, I watched all of it. It was very exciting with aircraft spinning down in smoke and people coming down in parachutes—I didn't know whether they were German or English. Those days will stick in my mind forever.

"I got into flying by accident," he said. In Britain the draft was extended after the war. "I elected to be conscripted for two years. One day they posted a notice about pilot training and I put my name in. I was lucky; out of 500, three of us were selected. After graduation came operational training in the Meteor, which was a hot aircraft in those times." During 23 years in fighters he served with squadrons based in England, Germany, and the Middle East, logging 4,300 hours in the Vampire, Venom, Hunter, F-86, F-4, and Gnat.

Loverseed instructed for four years at Central Flying School where the RAF has trained flight instructors for 75 years. As an exchange officer he instructed on F-4s at Davis Monthan AFB for two years. "Then I was lucky enough to be assigned to the Red Arrows for three years, the third as team leader. That was the peak of it all, the top job of all time. I couldn't believe it."

He prefaced everything with, "I was lucky . . . ." I am well enough acquainted with the Royal Air Force to know it is extremely fussy about staff assignments to CFS, representation in this country, and who joins its aerobatic team. Luck has little to do with it. Such coveted slots are filled by the best people there as they are here.

Surely he was headed for a rewarding military career? He smiled. "In the RAF, in order to get promoted, you do very little flying. In each rank you do one tour of flying, then three ground tours shuffling a lot of paper around. I preferred to fly. Besides, I was a poor little evacuee boy who didn't come up through the right schools and colleges." Again, military service there and here is more similar than different. In any case, an eye problem grounded him and he took medical retirement with the rank of squadron leader (USAF equivalent: major.)

"I preferred to fly." That sums him up. I've known many pilots who preferred flying to any job on the ground, no matter its rewards in money or career advancement. They were the finest pilots I knew. The art is advanced by them, not desk-flyers.

"I was u/s (unserviceable in RAF jargon) as far as the RAF was concerned. But the eye problem cleared up and I learned there were Hunters to be ferried to Singapore. I applied, without mentioning why I left the RAF, and got the job. That lasted two years. Then I was the demonstration pilot for DeHavilland in Canada for 10 years. That came to an end and I've been ferrying aircraft ever since."

He has ferried throughout the free world, logging 5,000 hours at it. He most enjoys the Far East—Thailand, China, the Phillipines, Malaysia; the least favorites, Libya and Iran. He's completed 120 Atlantic crossings in multiengine types; our Cessna would be his 11th in singles.

At Bangor, a 55-gallon tank would be installed on the rear seat to increase endurance to about 12 hours. "That will give me a ferry range of about 1,000 nautical miles with reserves," he said. An HF radio and battery-powered loran C would be secured in the right seat. I examined them, his immersion suit, life raft, and survival gear, and listened to his explanation of each. Interesting. With this seemingly jury-rigged gear he would tackle the broad Atlantic behind one little engine two men can lift.

"Are you apprehensive?" asked my wife.

"It depends on what you mean by 'apprehensive.' If you mean, do I look for the difficulties in it? Yes, I am. It's prudent to look at what could happen and to take reasonable preventive measures. If you mean am I fearful? The answer is no. If I thought it was daring, I wouldn't do it. Of course, if you've got to have cast-iron guarantees that you're going to be able to cope with anything that could happen, that's impossible."

Weeks later a neatly-typed letter from him arrived. He had flown the 4,500 statute miles in 41:10 with nine stops. From Bangor the route was Goose Bay, Newfoundland; Narssarssuaq, Greenland; Reykjavik, Iceland; Shannon, Ireland; and Bournemouth, England. After clearing customs, he continued to Goodwood airfield near Chichester, a Spitfire base during the Battle of Britain. Aside from being "bounced around uncomfortably" for two hours approaching Reykjavik, and being grounded there for a day due to high winds, the long journey was "uneventful," he wrote.

Footnote: Upon arrival, N9621H was hangared for removal of the ferrying equipment. That night the incredible unpredicted hurricane struck South England. Winds gusting to 100 mph blew off the hangar doors; five airplanes tied down outside were demolished, but little red-and-white Two One Hotel sustained no damage.

34

We can't tell it like it is

THE PURPLE PROSE IN MY MAGAZINE COLUMN is assembled with a word processor. It has a keyboard, TV screen, printer, and cannot process anything. I have to "process" my own words, that is, revise, correct, delete, insert, and transpose until the words on the screen seem right. The printer will transfer this column to paper in three minutes.

Then corrections are made—on the screen—and another copy is run off. The result is letter-perfect, no matter how it reads. It's wonderful insofar as the mechanics of composition are concerned. Once I got the hang of it, I sold my typewriter. Never miss it.

If you're thinking of getting one, be advised that computer salesmen don't speak English. The patter is about *bytes, Pascal language, disk drives, pathnames*, and *kernels*. I bought in spite of them and taught myself to use it. The handbooks supplied are pure gibberish.

It was like upgrading on the airline. The typical equipment manual provided a detailed mechanical description of the airplane. We had to memorize much of it, not in order to fly, but to pass the written and oral exams. It told you everything except how to fly the new plane.

Obviously, a pilot must have a good working knowledge of his aircraft to operate within its limits and cope with malfunctions. In the recip era we were force-fed such irrelevancies as tire pressures, strut extension, fundamentals of electricity, and the gauges of airframe dural. What this trivia had to do with operating the machine was never explained. "I want to fly one of these jobs, not build one!" groaned a patriarch in our DC-7 school. No matter, he had to recite dimensions and explain flush toilet operation along with the rest of us.

Moving up from the DC-3 to the Convair 340 or from the DC-6 to the Electra meant up to six weeks of intensive training, an absurd waste of time and money. The average candidate could have hacked it with two weeks of realistic instruction. Lecturers dwelled on petty details and flight training was unnecessarily complex and dangerous and caused the loss of an appalling number of airplanes and crews.

The introduction of jet equipment brought changes none of us expected. Emphasis on technical fine points was gradually replaced by a more operational approach. The new theme: If it fails, what can you do about it? Compared with the ordeal of DC-4 transition 25 years earlier, the 747 check out was a piece of cake. So accustomed had we become to intensive study of nonessentials, we wondered if we were getting enough background. We were. Midnight circling approaches at 400 feet with two shut down were now practiced in the simulator where clumsiness was embarrassing, not disastrous.

Surprisingly, this enlightened approach has not been adopted by everyone. FAA-approved down-to-earth curriculums are saving progressive managements considerable expense, yet others continue with costly pilot "training," which makes little sense.

An experienced DC-8 pilot can quickly transition to a DC-10 if his training concentrates on flying. There are some who enjoy learning much more about their airplane than is taught in school. I copiloted for captains who could describe in detail every component on board (and would at any excuse) but there was no obvious reason to regard them as better airmen.

An expert guess as to why an underfloor pump has failed is academic; knowing what to do about it is what counts. Indeed, preoccupation with detail can divert attention from the overall picture with terrible consequences. When trouble develops, keeping the airplane flying right side up at a safe speed and

Mexicana Comet 4C at Dallas on flight from Mexico City.

altitude is the prime concern. Dealing with a problem is of secondary importance. Many the perfectly flyable airplane that crashed while its crew fiddled with switches and recited checklists.

If you bought a computer like this one, I could in an hour teach you what it took me a month to learn—and you'd save all the paper I wasted. I'd give you a single sheet of do's, don'ts and reminders, all in plain English. You would then be able to write letters or themes, address envelopes, assemble records, and file away on disks anything worth saving. But don't ask me how it works. I haven't the foggiest; furthermore, I don't care.

While such total ignorance of the machinery won't do in piloting, there is a parallel. There is a Disney World aura about even the most realistic flight and/or simulator training. The modern simulator is a marvelous device for teaching normal and abnormal procedures and it provides an idea of how the airplane handles. In it we were "trained to proficiency," meaning until we could execute maneuvers and procedures within prescribed tolerances. When the examiner had checked all the squares on his sheet, we received the rating. This introduction did not provide all the answers, however.

It was common practice at our shop to ask an old head about operating a new type in the real world. What should I know about crosswind technique, icing, turbulence penetration, braking on slick runways, approach speeds, fuel consumption, high level cruise? What *didn't* I learn in school? The ensuing blunt advice was worth remembering.

We bought 720s to augment our jet fleet. They looked like 707s but were lighter and had smaller engines. Although sluggish by comparison, the new job handled enough like its big sister that a rating on one was good on both. There was one difference not emphasized during dual: when heavy, you'd best carry a few extra knots into the flare; otherwise, the 720 would arrive with a bone-jarring crunch as if you'd never raised the nose.

On his first 720 trip a senior skipper left half his main gear on the runway numbers and was cited for "reckless operation," although the flight recorder confirmed his speed at within two knots of the chart number. After much argument, the charge was dropped. The rest of us took note.

We also learned by experience and from each other that, when departing in summer heat or from a high altitude airport, it was unwise to haul the nose up promptly upon reaching V-rotate speed, no matter what they taught. Runway length permitting, you should pick up another 10 or so knots and then she'd not only fly, but climb. One down-on-the-deck mushing dash across the prairie leaving Denver made the point.

"Why can't the English learn to speak?" lamented Professor Higgins. And why can't we drop all the inane and superfluous double talk we use in aviation and speak plain English? If only we could discard all the manuals, regulations, procedures and policies and start over with a single rule: Tell it like it is!

A preposterous notion, of course. The whole idea of flying has become more important than the people who make it work. We say it serves us; in fact, we serve it. The system is demanding and unforgiving. Any group that does not measure up vanishes; any individual who through negligence or misfortune embarrasses the rest of us is cast aside, sometimes unfairly. All of which has made aviation like other orders of modern society, though its standards are higher than most.

The penalties for mistakes being what they are, today's pilot must do many things two ways—the book way and the way that makes sense—and know precisely how far he can bend the rules. Were he to comply in every instance with the letter rather than the spirit of the law, the system would collapse and everyone knows it. Yesterday's pilot was concerned for his life; today's is concerned for his job. There's far more to it now than flying skill.

It's a fair guess that many of yesterday's tough-minded individualists would not comfortably fit into today's hectic atmosphere. And that many of today's newcomers would not have found what they wanted in the way things used to be.

35

Training

THE EVOLUTION OF PILOT TRAINING since Kitty Hawk is as interesting as the evolvement of flying itself. In the beginning, the pioneering aeronaut learned from trial, error, and the often tragic blunders of his colleagues: no manuals or schools. The art of flying was self-taught.

The typical World War I flying school provided good ground training. The textbooks issued then are worthwhile reading today. The emphasis was on theory rather than practice, however. After weeks of lectures on navigation, engines, and armaments came flight instruction that was slapdash by comparison. A graduate could be sent to a fighting unit with 25 or fewer hours in his log, most of it solo time. Such a background resulted in a life expectancy of three weeks for replacements reaching the front during periods of intense activity.

The airplane was taught as a machine to be understood rather than a weapon to be wielded. I chanced upon the ground school notes of a 1917 trainee. His painstakingly-drawn illustrations were evidence that airframe, powerplant, and guns were studied in great detail. Intricate sketches of oil pumps and magnetos showed all parts identified. Photographs of the time show cadets rigging airframes and repairing engines. Why such thorough class work was not followed by equally intensive flight instruction is puzzling. (Trainee 2/Lt. A.J. Barbour, RFC: Did you survive in spite of it?)

A more realistic balance was struck between class and cockpit instruction in World War II. Two to three hundred hours, some logged flying operational equipment in simulated combat, were the preparation for war duty. Ground schooling followed the technical format established two decades earlier. I read several pages of the RFC subaltern's notes before realizing he had trained to fly Nieuports, not Spitfires. The basics had changed little between the wars.

The complexity of World War II aircraft necessitated extensive ground schooling, yet technical detail again was carried to an extreme. Like first war students, we dissected oil pumps, altimeters, and magnetos—along with such newer components as feathering props, superchargers, retractable wheels, and flaps. We wondered if knowing how a generator worked would help when it didn't, but failing meant washing out, so we memorized columns of irrelevant numbers that were in the aircraft manual if needed.

The command pilot of a multi-engine type was trained to perform, even in emergencies, as though his copilot and crew chief had stayed on the ground. In practice, many left-seaters made the work a team effort, yet the one-man-show doctrine prevailed simply because things had always been done that way. Years would pass before ground and flight training were made more realistic. In the meantime, operation in real-world situations largely remained a skill learned on the job.

The airline approach to training was a copy of the military, abbreviated for economy. Flight crew qualification was, and is, hugely expensive. The main aim was compliance with the law, no more nor less. Captain trainees were coached to pass federal written, oral, and flight tests. During the first postwar years new hires received virtually no training. Six of us were hired in the morning, given our CAA-mandated "three bounces" and sent out on the line that evening, thereafter to absorb wisdom by osmosis. The steadily improving safety record was a credit to individual self-discipline.

The DC-6 and Constellation changed the picture. Pressurization, synchronized reversible pitch props, two-stage blowers and water injection were among the new components to be mastered. Copilots received the same class work as captains; flight engineers endured even more intensive studies.

Classes were taught by mechanics; the material was technical and so elaborately detailed that one old-timer moaned in desperation, "I want to fly this airplane, not build one!" It was like requiring a driver to explain octane ratings and front- end alignment. The nuts-and-bolts approach was continued through the Electra, the ground schooling for which was an ordeal, to the DC-8 and 707. Lighted schematics of hydraulic and electric systems helped explain wiring and plumbing mazes. Then, to our amazement, the emphasis on mechanics began to wane.

Our line bought a 747 to fly the Texas-to-Hawaii run. Thinking it would be my only shot at the jumbo, I gave up my 707 captaincy for copiloting and went to United Air Line's school in Denver, dreading the worst. Surely, the monster's schooling would be horrendous. Amazingly, the approach there was cockpit oriented. Don't worry about how pumps and solenoids work, they said, only about what you can do from your seat up front when they fail.

We looked at each other in disbelief. Classes included slide shows with audio taped descriptions, after which an instructor answered questions. We were given exactly what was needed to pass the exams, no more nor less. In fact, when a question was asked about a point not covered, the answer was, "This course is FAA-approved. If it's not mentioned, you don't need to know." The final was a piece of cake.

Mornings in class was followed by afternoons in a cockpit procedures trainer (CPT), a true-to-life replica down to ash trays. Learn the layout, start the engines, deal with malfunctions, do everything but "fly." Simulator rent was $60 a minute (and that was more than 20 years ago); it was important to feel at home when you got to it. Flight training was moving indoors.

Ground school visual/audio aids with instructor backup caught on. In 1983, when I visited United's Boeing 767 school, there were no instructors in sight. Students sat in individual cubicles equipped with television sets on which cockpit panels were projected. Touch a switch or gauge on the screen and a system diagram appeared while a taped explanation played, a clever show repeated upon request. Ground school had become almost completely self-taught.

This radical approach eliminated the need (and expense) of full-time instructors, but was it effective? Some of the students I met were not enthusiastic. "I'm not sure I'm getting enough of this," said one. Another said the absence of an instructor to stress important points gave the syllabus an annoying sameness. "I learn from the questions of other students and class discussions and we have none of that." The school director said, "They do as well on tests as you and I under the old system. It all comes together in the simulator."

While teaching techniques underwent radical changes, the thrust of pilot training was still *get the ticket the quickest way*: Pass the exams, do the steep turns, stalls and engine-out approaches, recite the emergency checklists, get the rating. This produced technically sharp pilots, but the proper use of newly acquired skills in the real world of low ceilings and slick runways largely remained self-taught.

Confusion, lack of crew coordination, laxity and misuse of command authority continue to bring presumably qualified pilots to grief. While their misfortunes are put down as pilot error, their accidents clearly reflect on the quality of training.

In recent years, crew resources management (CRM) has been added to air safety lexicon. Progressive airline and business aircraft operators increasingly employ line oriented flight training (LOFT) in initial and recurrent training to foster optimum cockpit team work. A simulator LOFT "flight" is planned and executed from takeoff to landing just like a routine schedule. One or more abnormalities occur en route and diversion to an alternate terminal is probable. The exercise is videotaped, reviewed and, if necessary, repeated.

Such sensible expansion of training programs is sure to result in better cockpit performance. A little more time and a few more dollars spent on realistic crew qualification are certain to save lives and avoid monstrous damages in the long run.

36

Behind the cockpit door

TODAY'S AIRLINE PILOT IS ISOLATED from his passengers. On some trips he never sees them at all. He's missing something. It was different in the relaxed decade after World War II, before the air travel boom began. Airliners were much smaller, we flew slower, and traffic was light. There was time to mix with the fares, answer questions, reassure the nervous and get to know them, in fact, such glad-handing was encouraged by management. This good-will chore normally fell to the captain, of course.

We often learned who our riders were and why they were flying. Much of the public still regarded air travel as novel, even daring. Most passengers at that time were businessmen. Situations involving sorrow or good fortune sometimes accounted for other tickets sold. One example is unforgettable.

This story is incomplete because I was only there at the beginning. Actually, here are three tales and your guesses as to their endings are as good as mine. More than 30 years have passed since that summer evening trip, yet the notes I made jog a fair recollection of the details. Here is what happened.

We were working a Convair milk run that connected at Dallas with DC-6 nonstops from New York and Chicago and delivered through passengers to Wichita Falls, Lubbock, and Amarillo. After we had run the checklist, I went back to secure the commissary door, one of my copilot chores. There was commotion in the rear cabin as a harried woman attempted to get nine lively children settled in their seats. All wore Swissair tags. I wondered when and where their long day had begun.

They were quite young, the smallest being a tiny girl already sound asleep, and about evenly divided between boys and girls. The oldest was a blond-headed boy about five. He sat quietly and studied me intently with grave blue eyes. I offered him chewing gum and he carefully took a piece. On impulse I pressed the whole box into his hand and he accepted it without a word or change of expression. He was neither frightened nor friendly and I could only guess at what he was thinking.

Retired from airline service, this old Connie 1049 flew freight.

"Good load tonight," I told the captain, meaning most seats were taken. We junior pilots monitored load factors as closely as vice presidents for when they dipped, we (not the VPs) got furloughed. After takeoff the hostess came up with coffee.

"They are orphans who have been adopted by American families," she said. None could speak English, not even their guardian. The children had left Germany 20 hours earlier and would complete their long trip at Lubbock. A doctor and his wife became interested in war orphans and had arranged an adoption. Friends heard of the plan and more letters were written with the result that nine weary and bewildered youngsters were now speeding westward to new homes.

We landed about sundown. A large crowd was waiting behind the fence at the terminal. Several young men came out to help unload the small passengers. The scene that followed answered all questions as babies and toddlers met their new parents.

"Here she is," said one young man, handing a tiny girl to his wife. "Isn't she a little beauty!" The young woman hugged the child.

Another held a chubby boy at arm's length and his wife said, "I'd have known him anywhere. Look at him, look at that smile, will you!" and a slender boy in blue jeans said, "Let me see him, Dad. Here, let me carry him." The new arrivals were passed from parents to grandparents to uncles and aunts. Some of the foster parents were smiling and some were crying and all were very happy. Their emotions were infectious. You couldn't watch those selfless folks without feeling good and wishing this sort of thing happened more often.

"Just like her picture, she's a doll, exactly like her picture," someone kept saying.

"Look, he's so tired, poor little thing, we must take him straight home to bed," said an older woman, busily settling into her new grandmother role. But first there were pictures to be taken for the papers and interviews with the new parents. Then the crowd broke up into small groups heading for their cars. I looked for the little boy who got my gum. He was with a young couple, as quiet as he had been on the airplane, looking from one to the other with the same grave inquiring expression.

The woman fussed with the buttons of his coat but little was being said. The man picked up his small suitcase and the three walked toward the door, the boy between them holding their hands. They did not seem like people who talk much and I imagined the boy had found his way into exactly the right home. I hoped so and was glad to have had a hand in his finding it. We Americans take too much for granted.

We headed north toward Amarillo. "Wonderful thing, you know," said the captain. "Those kids coming way out here to grow up, think on that. What a change from what they left, not that they'll remember much about it. And those folks who took them in. Good people."

The hostess came up with more coffee. "Did you notice the lady up front with two young girls?" We had; she was an attractive young woman about 30 with two pretty daughters about six and eight. "I've been talking with her. She's had it rough. Her husband had been gone a year. He was an Air Force pilot who was killed three weeks ago in an accident in England."

You'd never have known it from seeing her, we said. "No, you wouldn't. She's completely composed. She said, 'The girls and I know their daddy isn't coming home again and that we are going to have to stick together and make a new life without him.' "

"Just like that?" said the captain.

"Her exact words. You know, I don't think I could take it like that."

"She's cut from tough cloth," he said, "like the people who opened up this part of the country not so many years ago."

"She said she was born and raised in Amarillo."

"I can believe it."

The last red streaks of daylight were suddenly gone and darkness creeping up from the east made the night complete. The dim glow of our destination lay over the nose. It was always interesting to learn about the souls behind us. They came aboard, we flew together, they walked off, and out of our lives. For a brief time we shared the world unto itself that was an airliner in flight. Who were they? Where were they going? During that special evening we learned some of the answers.

"Don't forget the package," said the captain, meaning the white metal canister behind his seat. It was a metal can about the size that holds a pound of pipe tobacco. An agent brought it aboard in Dallas. You will be met by a doctor, he said. Please see that he gets it right away.

The routing tag said it came from an eye bank in Houston and was en route to an Amarillo hospital. It was unsealed. Inside, two little transparent jars floated in a pool of ice water. They were half-filled with cotton and on each swab lay an eye. They were brown and, according to the label, had given sight to a living person earlier that same day.

A doctor met us in operations to collect the package. He took time to answer questions. Cornea transplants would be accomplished that night; it was a routine procedure, he said, and the chances of success were excellent. He hurried off; we prepared for the return trip.

I can't give you the rest of those stories, Paul Harvey-style, but I make these guesses: the children grew up to be solid citizen Texans. They are in their late 30s now and might be taking their own kids to see Germany. The Air Force widow eventually remarried and is now a grandmother. On a certain day each year she looks at a picture mailed long ago by a pilot in England. And that midnight eye operation? Of course it was a complete success. Who could think otherwise?

As our passengers boarded for the return trip to Dallas, the captain pointed to a couple behind the fence locked in passionate embrace; it was a steamy scene. The girl, a stunning redhead, broke away and trotted toward the airplane, dabbing her eyes. She was the first off at Lubbock where she ran into the arms of another young man and kissed him with apparent deep feeling.

The end of that story? Your guess is as good as mine.

37

Seniority

YOU WATCH THEM BOARD A 727: pilot, copilot, and flight engineer; in airline parlance, captain, first officer, and second officer. Because they seem about the same age, a layman might wonder why one has four stripes and scrambled eggs while the other two wear fewer stripes and plain-bill caps.

Did the captain have more college credits or the most logged time? Did he score the highest in ground school? Did he know someone in management? Did he marry the chief pilot's daughter?

The answer is none of the above. He's number one in the crew simply by being hired before the copilot who, in turn, was hired before the engineer. An impressive resume or pull with a company official might have helped land the job, but he began at the bottom with everyone else. Length of service put him in the left seat, the copilot in the right, and the engineer on the panel behind them, and they are locked into their relative positions for as long as they work for their line.

Airline pilots work within the seniority system, which is often derided by those within a "merit" system. The latter say the seniority system ensures promotion with no regard of individual initiative, while merit system workers move up if they perform, and down, or back out the door, if they don't. At least, that's the theory. In practice, merit system underlings advance when they please their overlings, which often leads to what is politely known as brown-nosing.

But not to debate the issue. Each way has its points, good and bad. For 60 years the seniority system, as adopted from railroaders, has been the airline way and it is not apt to be changed. For anyone with airline piloting aspirations, understanding and accepting it is as important as building time.

Recently a young pilot wrote to me. He's logged 1,500 hours, 500 in twin turboprops, enough to interest a major carrier: "First, I'm going to complete college and get my degree, then I'll apply." If you have your heart set on airline

The Boeing 707-320C "Intercontinental" was a pilot's dream to fly.

flying, listen carefully; that lad doesn't have the picture. Get your name on a seniority list the quickest way. Shelve everything else until you've nailed down that all-important number.

Let's say Trans-Parent Airlines is hiring 40 pilots a month. You are accepted and asked if you want to report now or for a later class. You have personal matters to attend to, so you elect the third class. You have really screwed up. The 80 pilots hired while you dawdle will remain ahead of you forever, meaning they will fly nicer schedules in better equipment, live in more desirable bases, sweat fewer furloughs and earn more than you ever will. You will never overtake a single one, even if you're better qualified than all of them.

If you think 80 numbers are not worth worrying about on a 3,000-pilot roster, hear someone who's been there. At an early stage, 25 numbers lower would have put me on extended furlough. My long-awaited checkout as captain was six months behind the fellow five numbers my senior. One number put an interesting international schedule beyond reach. Three numbers avoided an undesirable base assignment. Those circumstances are recalled decades later. A single number can cost a pilot many thousands in income over the years plus many other good things. Do I exaggerate? Ask any other airline pilot.

Of course, such was the lot of all of us. Seniority was—and is—everything. Number one is the only pilot who has it all his own way. Number two must settle for second best, number 100 gets what his 99 seniors leave and so on. The top numbers enjoy the feast, the lowest scramble for the crumbs.

You may envy a senior his number but begrudging him it would be like begrudging an older brother his date of birth. Politics aren't in the picture, a seniority system plus. You'd not want to ride behind a skipper aware that the copilot was after his seat—and might get it.

You advance as a group in good times and sink together in bad, the bottom numbers being pink-slipped when load factors dip. Sometimes the picture is rosy, when business is booming, when hiring is up, when hundreds of new airliners are on order, and when there's a pilot shortage. But the past is prologue; the pendulum will swing. I'm not a pessimist, but a realist thinking back. Get on a list now and pray that the "pad" building beneath you will cover the layoffs future retrenchments will bring.

The seniority system guarantees a chance at promotion, not promotion itself. The new hire who imagines his stint as flight engineer automatically leads to a copilot's slot will learn differently. First, there's groundschool and simulator/flight training during which his background counts for naught. He advances only if he cuts the mustard. If he cannot? That depends; on some lines he resumes his third seat pending another attempt, or indefinitely if he chooses. There are by choice a number of "career" engineers (and copilots.) Other carriers demand upgrading; failure to qualify means dismissal.

Whatever its failings, the regulated era lent the work a feeling of stability. The company depended on us, and we on it. We took pride in a career most people respected and many envied. We knew where we were going and, excepting severe economic slides, when we would get there. We looked to the future with confidence. Attrition and expansion steadily, if slowly, lifted us toward better things.

For many pilots that agreeable outlook has been replaced by a dread of what's coming next. Bankruptcies, mergers, and takeovers have thrown the profession into turmoil. A buy out entails far more than agreeing on price. Assume Trans-Parent, whose junior captains have five years of service, buys Trans-Sonic, whose junior captains are 10-year people. Do you merge the captains of both lines? That's the "relative position" answer. Or do you go by date-of-hire, regardless of crew position? What's fair? What's right? It's a negotiator's nightmare.

A pilot regards any fiddling with his precious number as he would a pass at his wife. Whatever the mix, it is certain to be greeted with howls of dismay by many. When my company acquired a smaller line long ago, the merged list worked out after heated debate produced bitterness in both camps. The grudge was cemented in some cases. "And it will last until we're all retired," I said to a friend.

"No, it will last until we're all dead," he said.

In some cases the job security known prior to the revolution suddenly evaporated. Braniff's collapse in 1982 idled 1,200 pilots with little warning. At first it seemed the end of the world. The youngest would find other berths but what about those 35 and over? Traditionally, if you weren't established by age 30, you could forget airline flying.

But, whereas deregulation meant ruin for one line it spelled opportunity for another; the survivors were soon short of pilots. Age limits were forgotten and displaced 40-year olds happily found themselves in demand. Today Braniff veterans fly with a score of airlines. Seniority is not transferable so it was back to the beginning for those hired by other lines.

Individual seniority was affected in strange ways. After 17 years with Braniff, one captain was, in round numbers, Number 325 out of 1,200. After seven years with Piedmont he was 715 out of 2,525 and the merger with USAir puts him 2,300 out of 5,630, thus he went from the top 27 percent to 28 percent to 41 percent. A younger pilot who was 54 percent with Braniff now ranks 40 percent after the Piedmont/US Air merger. Both of them are now ahead of Braniff friends hired later, some of whom were once their seniors. Further upsets of former relative positions occurred because USAir had also been hiring ex-Braniff pilots.

None of which alters the importance of landing a job the quickest way. Many of today's new hires will not enjoy the relatively smooth career ride our generation knew. The turmoil has not ended, and like the weather, it is beyond the control of the individual pilot. The realistic outlook has to be: prepare for the worst, cinch your belt, and hope for the best. After all, isn't that what being a pilot is all about?

38

The curse of change

THE ONLY WAY TO LEARN a new airplane is to fly it for as long as it takes to become completely comfortable in all situations. The private pilot is fortunate; he can pick the time and place to try crosswind arrivals, short field operation, night, and instrument flying. An airline pilot learns on the job with 50 to 500 trusting souls in back. Concentrated ground schooling and simulator practice cannot answer all the questions that arise during all-weather flying. As the old saw goes, a type rating is only a license to learn.

It's also said that a new captain will encounter the gamut of grief on his first trip. That certainly was my fate. You name it, we ran into it during three days in that old Convair—ice, gusting crosswinds, slick runways, tight approaches, and a go-around when the trip ahead blew a tire. At least I had a sharp copilot.

My first month on the 707 included a round-trip from Dallas to Minneapolis with a stop each way at old Kansas City Municipal. It was March and the snow melted by day and froze at night. The wind was invariably from the west, which was 90-degrees off the only runway. So it was a crosswind arrival with "braking action reported as poor to nil" time after time. Again I drew a fine copilot but, alas, he was as green as I. Somehow we managed.

A newly-qualified pilot knows more than he realizes but he has to prove it to himself. Time and sweat are required to get ahead, and stay ahead, of a new type and reach the point where it no longer surprises. In the meantime, its alien sights, sounds, and responses make him uneasy. In flying, familiarity breeds competence.

One spring afternoon, soon after we received our first jets, a solid line of thunderstorms developed south of Dallas. A senior skipper heading for Houston gazed at the evil clouds and his spanking new 707 parked outside. He turned to the rest of us and said, half-jokingly, "I'll pay $500 to anyone who will fly this trip for me."

Abandoned Cessna C-78 at Denver was only good for parts.

It is the nature of our game that a pilot must continuously adjust to new rules, procedures, ground facilities, and cockpit equipment. And, several times during his career, he must upgrade to a different and more demanding airplane. Understandably, he comes to detest change. It's not just learning new aircraft; his company mail box stays filled with revisions, admonitions, warnings and reminders, most of it pettifoggery conceived by ground people with nothing better to do, but every item must be read. Change for its own sake is a curse of the trade.

Qualifying on a strange airplane is even more worrisome when it's a new design. By the early 1950s our trusty DC-3s were ending their trunk line usefulness. We knew they would go, but an order for Convair 340s made us wonder. What did Convair know about airliners? They built B-36s. Douglas and Lockheed built airliners.

According to the specs, the 340 would be faster, pressurized, haul twice the load, and feature reverse-pitch propellers. A pilot sent to the factory returned with a bizarre report. If you lost an engine on takeoff, he said, you didn't "bend it over and clean it up," DC-3 fashion, but kept climbing with the flaps down; otherwise, it settled. And the failed engine's prop feathered itself, if you could believe it.

We soon found the 340 to be a splendid airplane, however. Yes, it would climb with an engine out and flaps down and yes, it would sink if you pulled them up. If the auto-feather circuit went haywire and tried to stop a good engine, you simply pulled the feather button back out. Few airliners entered service with fewer bugs. Convair was added to our list of trustworthy suppliers.

This was not always the case. A number of newcomers emerged with serious flaws, the DC-6 and Constellation among them, and no one involved can ever forget the Electra or Britain's Comet. All went into the books as great aircraft, which indeed they became after design mistakes were set right. An official stamp of approval does not mean that all experimental flying is finished. An airline pilot does not fancy himself a test pilot, yet, to a degree, he fills that role.

In fairness, a builder cannot simulate all the rigors of scheduled flying. If you want to give a car the ultimate test, sell it to a Manhattan cab driver; to assess a transport airplane, sell it to an airline. You will not be long finding out if either is properly nailed together.

The greatest change in airline piloting came with the switch to turbine power. We got our indoctrination in the Electra, the groundschooling for which was pure nightmare. Conversely, it was easy to fly and handled like a four-engine fighter. After two disastrous structural failures, it was modified and eventually earned a trustworthy reputation. The Electra's Allison turboprops were but a hint of what was to come.

The monstrous (it was then) 707 was preceded by horrendous rumors. Because its engines were suspended beneath the wings, they said you had to touch down in a wings-level crab in a crosswind or you'd drag a pod. With flaps retracted, outboard ailerons were locked in trail and banking accomplished with small inboard surfaces. Even idling thrust could flip a small airplane onto its back. It was reportedly so different from prop-driven types that many older pilots would find the adjustment beyond them.

With such horror stories in mind, our seniors viewed transition with trepidation. A few elected to finish their careers in DC-7s. Those who volunteered made the grade without exception. The big Boeing was certainly different and more demanding, but it could be mastered and what a dream come true once it was.

That some of the top numbers needed extra dual was not surprising. They had flown the mail in DH-4s and passengers in Ford Tri-Motors 35 years earlier. The advance from 90 mph in an open cockpit to 600 mph at 41,000 feet, no matter how gradually accomplished, required remarkable effort and self-discipline. Old pilots can be taught new tricks, but they will never like the teaching.

Retirement from the airline meant the loss of so many good things that I made it a point to remind myself of those things I didn't miss, like picking the way between thunderstorms on moonless nights, sweating fuel, marginal airports, checkrides— the list is long. Most of all I didn't miss the endless changes of every nit-picking detail of the work. It was a wonderful life right to the end, but the steadily increasing hassle that became part of it would drive a saint to drink.

No sir, retirement ain't all bad. I bought a little red-and-white Skyhawk and made friends with it on my own time. No cram courses, no multiple-choice exams, no orals, no check airmen, no pressure—just friendly advice from fellow private pilots who were surprised at my elemental questions. They could not appreciate the immense gulf between their world and the one I'd left. In due time I felt reasonably competent.

There was sufficient gear in N9621H to shoot approaches at O'Hare in December; it was curiously satisfying to leave it all turned off once clear of our rural field. For the first time in 45 years I flew without a schedule, destination, or economic purpose. Man and machine alone in an empty sky as it was in the beginning. Every retiree should have an airplane. When Two One Hotel was flown away to its new owner in England, I wondered if we'd find another airplane as nice—another Skyhawk, that is.

"Try something different," said a friend. He had a Grumman AA5B at the time, a sprightly performer with a 180-hp Lycoming. Grumman owners are smug in the belief they fly the best, which makes them no different from owners of Cessnas, Pipers, Beechs, Mooneys, and you name it. Then I tried a Piper Warrior and bought it. It came without intensive training and manuals revised every week. After three times around the pattern I was on my own to learn it at my own pace. *That* much change I can still take.

39

Benefit of the doubt

UNTIL HIS DEATH, I had never heard of Captain Will Rogers. What was said afterwards provoked a feeling of kinship with a pilot I never knew. He was one of us, I know that much.

Rogers, an American, flew for Surinaamse Luchvaart Maatschappij (Surinam Airways), the national airline of Surinam, a self-governing territory of The Netherlands on the northeast coast of South America. U.S. Army pilots who flew South Atlantic routes to Africa during World War II knew it as Dutch Guiana. Two DC-8s and two Twin Otters made up SLM's fleet.

One evening in June, 1989, an SLM DC-8 commanded by Rogers departed Amsterdam for Paramaribo, Surinam's capital, with 188 souls on board, arriving just after 4 a.m. Two unsuccessful approaches were made through "dense fog." The crew requested that the runway lights be turned up; there was no mention of an emergency. On the third try the aircraft struck a tree, breaking into four sections. Although another pilot saw "a ball of fire" prior to impact, there was no post-crash fire. Captain Rogers and his crew were not among the survivors. These are the facts as reported in newspapers and on television. After three days the crash received no further attention in this country.

Within 24 hours a "government spokesman" said, "We think it was pilot error." An SLM official added that technical problems had virtually been ruled out because the airplane had a maintenance check two weeks earlier. This finger-pointing could only rankle anyone with the least awareness of possible causes. A team of National Transportation Safety Board investigators had yet to reach the scene.

A typical NTSB investigation of an airline accident requires months of tedious analysis and while its findings might seem conclusive, the cause is *always* listed as "probable." Opinions circulated in the meantime are pure conjecture.

Speculation about causes must explore all possibilities. The Amsterdam-Paramaribo run is flown in about nine hours. Even under ideal conditions, such a night schedule is a grind. Poor terminal conditions at dawn are common in certain seasons and locations; dealing with them is part of the job. It is completed without incident many times every day. A mistake by a fatigued crew shooting a tight approach in the early hours? A possibility, but only one of many.

A malfunctioning navigation aid can mislead a crew as to its precise position during arrival. A similar accident in the United States and many other foreign countries would call for an immediate flight test of all terminal navaids. The NTSB probably requested such tests at Paramaribo. The wealth and reliability of aids and weather services we take for granted are not always found in other countries. According to one account, there is no radar equipment at Paramaribo. A trip into many foreign airfields is a trip into history.

Faulty flight instruments have led unsuspecting pilots into the ground. Yes, I know modern airlines have dual flight instrument groups, each drawing information from its own sources. I have seen both groups go haywire simultaneously (fortunately during cruise in clear skies).

Crew incapacitation is a possibility. A seizure or sudden illness experienced by pilot, copilot, or both during the final critical minutes could cause loss of control. Autopsies might determine if this was a factor. Was anyone besides the crew found in the cockpit wreckage? What conversation was picked up by the cockpit recorder?

Scrutiny of passenger lists has provided valuable clues in the past. Could one rider have been a hijacker or a madman bent on murder or suicide? The cargo manifest will be equally interesting. Toxic fumes leaking from hazardous material or a corrosive liquid dripping onto wiring or plumbing can have undetected lethal effects.

In the light of recent history, the specter of a bomb looms large. A detonator can be set to trigger when a certain level is reached during climb or descent or any time the vermin who assembled it wants his demoniac device to explode. While there is no suggestion that the political climate in Surinam is anything but tranquil, it is worth noting that its army chief of staff, chief of operations, and air force commander were aboard. The NTSB will learn what caused that ball of fire.

Recently a 727 hit a large bird that penetrated the radome and pressure bulkhead, smashed through a rudder pedal, and mangled the copilot's leg. Bird remains have been found in several wrecked cockpits. If the weather at Paramaribo was accurately reported, a bird strike seems unlikely, as does a lightning strike or wind shear. While "dense fog" suggests a low visibility, windless situation, these ideas cannot be ignored.

The absence of fire after impact could mean there was no fuel left to burn. Did the DC-8 arrive at Paramaribo with near-dry tanks due to faulty quantity gauges or a leak that occurred past the point of no return? If so, the two missed approaches quickly consumed much of what remained and subsequent maneuvering might have uncovered tank outlets, causing one or more flame outs. There are limits to climb attitudes and turns when tank quantities are low.

The NTSB people will determine if all engines were developing power at the time of impact.

They will also want a look at hydraulic fluid levels. Loss of fluid can render control boost systems inoperative. While a DC-8 can be kept flying without boosts, the procedures are extremely demanding. A boost-off instrument approach to a fog-bound runway is a nightmarish exercise. Time-consuming emergency extension of the landing gear and flaps could further complicate the dilemma.

A recent maintenance check does not guarantee a like-new airplane. In fact, airline pilots much prefer a ship due for a check to one just out of the shop. The first is a known quantity, the second a question. When mechanics disassemble a large airplane to check, repair, and replace its many components, there is considerable room for error, no matter how thorough their work and how often it is double-checked by inspectors. An occasional oversight is inevitable and expected.

The fact that Captain Rogers did not report a problem by radio is not proof that none existed. If trouble developed while circling for a third approach—smoke in the cockpit, loss of an engine, a control failure, whatever—the last thing he would reach for is the mike. After an accident, laymen assume that because no distress call was heard, everything was normal. A problem encountered during the final moments of letdown to a fog-bound runway demands complete crew attention. A number of actions have to be taken immediately and completed exactly right. During such emergencies, the mike is the most useless piece of equipment on board.

All of these possible scenarios have precedents. Many others can be cited: a malfunctioning electrically-heated windshield that suddenly becomes translucent at the worst possible time, asymmetrical flap extension resulting in an uncontrollable roll at low level, or a runaway stabilizer trim. Jammed flight controls and exploding oxygen bottles have caused accidents.

Out-of-limits load distribution, a cockpit or cabin fire, and complete loss of cockpit lighting have been the causes of others. Inadvertent engine thrust reversal, a fuel cell explosion, inaccurate navigation charts and instrument approach plates, a structural failure, a surface-to-air missile strike, contaminated fuel: The possibilities are almost endless and all must be considered.

Those of us who have "been there" reject the snap conclusion of pilot error. We have endured those black hours in Captain Rogers' seat and ended the interminable night with a ticklish approach. We were often low on fuel and now and then needed more than one shot to put the mains on the asphalt. A hundred things could have spoiled it every time.

No one yet knows what transpired during the final minutes of Captain Rogers' third approach. Unless and until pilot error is established as the probable cause, it is only fair and decent to give him and his crew the benefit of the doubt.

40

The A320 Airbus

COMPUTERS ARE A MYSTERY TO ME and I don't care to have them explained, to paraphrase James Thurber. They balance my checkbook, water my lawn, and regulate my car's ignition . . . all very wonderful, all very mysterious. Now there are computerized airplanes, the F-16 for example. Of course, its pilot can eject if the computer is "down" as they say when the screen suddenly goes blank.

At first blush, the idea of a computerized *airliner* is preposterous, but a "fly-by-wire" airliner is already in service—the Airbus A320—and doing famously. I was invited to fly it. When N901BN taxied in, there could be no doubt about its owners; BRANIFF in huge red capitals ran from cabin door to rear window. (This was not long before the resurrected Braniff collapsed for the second time.) The idling turbofans were surprisingly quiet even at 50 feet away.

From a distance the A320-200 (only 22 -100s were built) resembles the Boeing 737-200, but in size and weight it is more like the McDonnell-Douglas MD-87. It is formally described as a fly-by-wire transport with high-bypass turbofan engines, plastic tail units, full-authority engine controls, integrated digital avionics and cockpit displays, carbon brakes, radial tires, and a new wing design with tip fences.

The A320-200 accommodates 150 passengers, 12 in first class. It is powered by two International Aero Engines (IAE) V2500 turbofans assembled in England from parts made in the USA, Great Britain, Japan, Germany, and Italy. The V2500 is rated at 25,000 pounds of thrust.

The purpose of our six-hour hop was to complete the rating rides of Captains Jerry Wolfe, Pete Krause, and Chuck Greene, copilots of mine years ago. We took off just before midnight and flew to a deserted airfield near Miami. With 40,000 pounds of fuel in the three tanks, we weighed 120,515 pounds, 42,525 pounds below maximum takeoff weight.

While the trainees demonstrated their skills to an FAA inspector, I explored. The cabin is eight inches wider than a 727's. You can slip by the cart in the A320's aisle, a comfort for the poor soul whose urgent need of the facilities invariably coincides with meal service. It's a quiet airplane. Oddly, the quietest seats during takeoff are the last row in coach. The overhead bins are enormous. The seats are of first quality, the leg room more than ample. Passengers will like the A320.

But they won't applaud after landing. The new Airbus gives its pilot even less than the stiff-legged Electra. I rode through a dozen arrivals, all of which rattled the galley equipment. Its firm return to terra firma is no discredit to the fellow on the stick. "It'll humble you," said Jerry.

What did my friends think of flying-by-wire? "I love it," said Chuck. "Once you understand the concept you have complete faith in it," said Pete. Jerry was equally enthusiastic; he sounded like an Airbus salesman rather than a newly-trained pilot. Mind you, these are not computer age youngsters but 35-year veterans typically skeptical of anything new or different. Their enthusiasm surprised me.

What they didn't like was the training. "It was a bitch," said Chuck. Something like the transition from reciprocating power to turbines back when we got the Electras and 707s? "Worse than that. This fly-by-wire business is different from anything we had in the past." There is much to learn. The average simulator rating ride runs three hours. That also must be a bitch.

The fly-by-wire system is a logical application of the latest technology. In the beginning the pilot's control wheel (or stick) and foot pedals were directly connected to ailerons, elevator, and rudder by cables and rods. Trim tabs were later added to neutralize control forces induced by changing speed and load distribution. This arrangement worked nicely through types as large as the DC-7 and as fast as the P-51.

Dramatic increases in speeds and weights after World War II led to hydraulic boosting—similar to power steering—and electric trimming. When one or both failed, manual reversion procedures, same requiring the combined muscle of both pilots, were used to maintain control. At high speeds, boosted controls made overstressing of the airframe a dangerous possibility. Feel sensors were installed to moderate booster strength as speed increased, duplicating the stiffening of controls felt in nonboosted types.

The jumbos posed new problems. No two Steeler linemen could budge a 747's controls on the ramp, much less in flight, without assistance. There would be no manual reversion if boosters failed. Redundancy was the only solution. Boeing's giant has four independent hydraulic systems, any one of which can drive all control boosters.

The engineering of flight and engine controls is tedious and expensive. A friend who worked on a fighter design spent two years devising the linkage between throttles and engines. Wiring can be routed through an airframe more easily than cables and rods; the possible savings in design costs, construction time, and weight are obvious. Fly-by-wire was inevitable.

Operating components "by wire" is not a novel idea. The 707 appeared in 1954 with motorized fuel valves; twist knobs to open and close. Old-timers were dubious: "A short circuit would leave me with fuel I can't use," said one. The new way proved entirely reliable, however. If fuel valves could be electrically positioned, why not flight controls?

But there's more to flying-by-wire than electric activation of control boosters. "Fly-through-computer" better describes the system. The A320's stick sends signals to SEVEN computers and *they* control the boosters. This family of electronic wizards shares the work, monitors one another, and automatically assumes the duties of an ailing member, any glitch being displayed on the panel along with advice as to proper crew reaction.

The A320 is not the first fly-by-wire airplane, or airliner. The Concorde has been flown-by-wire for 20 years. Its mechanical backup system has yet to be used, according to Airbus representative Dave Hopkins. General Dynamic's F-16 is the first fly-by-wire production fighter and Airbus' earlier A310 had fly-by-wire flaps and slats. The A320 is the first airliner with fly-by-wire *primary* flight controls.

Response to stick inputs is not blindly slavish, the computers having been programmed to respect all operating limits. A pilot cannot exceed them. If, due to fatigue, distraction, or oversight he commands a speed or maneuver outside the normal performance envelope, he will not get it; thus, he cannot stall the airplane, overspeed, or otherwise stress the airframe. In effect, the A320 takes care of itself while protecting the pilot from himself.

Power management through computers results in more miles per gallon than the most conscientious pilot could obtain by manually adjusting the thrust levers. With an autopilot engaged, the airplane is continuously in trim, exactly on course and speed, and burning the exactly right amount of fuel. The weight saving of 800 pounds through the use of computers and side sticks translates into additional payload, smoother flight, and lower operating expenses.

All of which is fine and dandy so long as everything works. My recollections of airline flying are that electric and electronic components were more prone to malfunction than those operated mechanically and hydraulically. Ground school obviously erased the skepticism of my friends. They are clearly convinced the overlapping of computer functions and other safeguards produces acceptable redundancy. A detailed explanation of the system would likely allay my own doubts. We fear what we don't understand.

According to Hopkins, the Concorde has yet to sustain complete computer breakdown. He says a test-flown A320 took numerous lightning strikes without any effect on its computerized controls. A single generator or the auxiliary power unit can supply power for normal operation, with a wind-driven generator and two batteries as backups. Studies have shown that the chance of complete computer failure is about that of losing a wing. The validity of this conclusion can only be tested by time.

What if all computers *did* fail for whatever far-fetched reasons? The crew would be left with the rudder, which is not fly-by-wire, and stabilizer trim. These will provide control until computers are reset. Test pilots have landed in this mode but the procedure is not taught nor required on rating rides.

"Take us home," said Captain Wendell Stephens when the night's work was done. Wendell and I go back to 1957 when he warmed the flight engineer's seat between the captain and me in a Braniff DC-7C. He was manager of A320 training with the "new Braniff," which had resumed flying in 1984 after its bankruptcy two years earlier.

The pastel blue cockpit was roomy, uncluttered and inviting. There was an instinctive urge to bring the new wonder to life. The captain's seat motored forward at the touch of a button. The wide armrest positioned my grip on the side stick, which immediately felt natural, although the absence of a control wheel and column was a glaring omission. The panel-mounted six large cathode ray tubes displayed a bewildering array of multicolored symbols and numbers. It was like sitting down to play the world's greatest video game. "Any kid sharp at PacMan could fly this airplane as well the first hour as we could after 25," joked Wendell..

Firing up is easy: fuel control switch ON, sit back and watch. The engine spools up, lights off, and any tendency to hang or overtemp is recognized by the system and dealt with automatically. The idling engines can scarcely be heard above the muted hissing of air from the vents. It was a muggy 93 degrees on the ramp but a comfortable 70 indoors.

Brakes off. Nosewheel steering was sensitive and required a watchmaker's touch during turns. A button momentarily immobilized the system, allowing the rudder to be checked full throw while taxiing. The appropriate numbers and codes having been loaded into the computers, we were ready to roll upon reaching the runway. When Wendell said, "You've got it," I pushed the thrust levers forward to the proper notch. There's no concern about over-boosting the engines (you can't) or fine tuning (no need to). Having considered temperature, runway length and weight, the computers ordered up the precise fuel flow for takeoff thrust. Engine noise at full thrust was a deep drone I remember from 747 flying.

Acceleration was brisk; the A320 wanted to fly. Keeping nosewheels on the centerline with the rudder pedals was a snap. Although the computer-calculated V-speeds were displayed on the flight director, Wendell called them out. At V-rotate I eased back on the stick, the nose rose 12 degrees and the rate-of-climb pegged out. Noise abatement fans will love this one. "At max gross takeoff weight we could lose an engine now and still climb to 17,000 feet," he said.

With gear and flaps stowed, I pulled the thrust levers back to the CL (climb) notch. There would be no need to touch them again until the flare for landing, nor would any retrimming be required. Set trim before takeoff and forget it. Fuel management, including cross feeding, is fully automatic, as are air conditioning, hydraulics, and electrics. Malfunctions are reported on the panel along with the correct pilot response.

A blue vertical line on the flight director indicates course, a horizontal twin shows vertical displacement. They work like localizer and glide slope needles on early ILS gauges. A small yellow square floats behind them. Keep the blue lines crossed over the square and you're on the programmed flight path. This required concentration but they say it soon becomes second nature. Subtle nudges of the stick brought immediate response. You almost think your way through turns. Speed and altitude are presented on vertical tapes that seemed too small, or perhaps I needed new glasses.

At 10,000 feet, the horizontal line slowly dipped, requiring forward stick to recenter the square. Of course, for we were clear of the 250 knot speed limit. Soon after, following another dip smoothly leveled us at 22,000. Fuel flow stabilized at 2,500 pounds per hour on each side, about half a 727's consumption. Wind noise at Mach .78 was minimal.

All too soon the blue line dipped, calling for descent. Orlando lay on the dark horizon. Keeping the yellow square centered leveled us at 10,000 feet and kept speed at 250 knots thereafter. I picked up the VASI lights and aligned us with the runway, stick corrections by then having become almost instinctive. Who needs a wheel? A manufactured male voice called out 200 feet, 100, 50 and then, "Retard, retard, retard . . . ," meaning to ensure the thrust levers were back to idle.

I will not dwell on the touchdown except to say again that the A320 is stiff-legged and it had been seven years since I flew anything heavier than 2,500 pounds. After securing, a cash register-like tape slowly rolled out of the pedestal. It was a complete record of equipment performance for the flight, invaluable to the maintenance people for trouble-shooting.

Airbus Industries is as interesting as its product. With headquarters and assembly plant in France, it is often considered a French company. In fact, it is a multinational consortium created in 1970 as a "Groupement d'Interet Economique," a legal entity enabling members to work together on group projects while pursuing other projects independently. Its risk/profit sharing partners are Aerospatiale of France (37.9 percent), Deutsche Airbus of West Germany (37.9 percent), British Aerospace of Great Britain (20 percent), and CASA of Spain (4.2 percent). Fokker (the Netherlands) and Belairbus (Belgium) are associate partners. The combined cooperative assets are $11.9 billion.

Airbus is by no means an industry upstart. Its family tree is a veritable roster of famed European builders, among them Nieuport, Caudron, Farman, Focke-Wulf, Messerschmitt, Junkers, Avro, De Havilland and Bristol—impressive bloodlines indeed. Several of its ancestor firms were building airplanes and engines before the first of its American competitors came into being.

Its first design, the 267-seat twin-engine A300, entered airline service in 1974. The shortened A310 (218 seats) followed in 1983. The A330 (328) and A340, the latter a four-engine model seating up to 440 coach passengers, are currently in production. The revolutionary A320 began airline service in 1988 and there are already orders in hand for a stretched model, the A321.

More than 60 European plants employing 28,000 workers are engaged in AI projects. Some Deutsche Airbus work is subcontracted to Italian firms, some British Aerospace work to the Australian industry. In addition, up to 30 percent of an AI aircraft consists of American-made components.

The consortium functions like this to produce an A320: Aerospatiale builds the entire front fuselage, cabin and nosewheel doors, center wing box, engine pylons, and is responsible for final assembly. The center and rear fuselage, wing flaps, fin, and rudder are built in West Germany. British Aerospace builds the wings, including ailerons and spoilers and the gear leg fairings. Belairbus produces the leading edge slats and CASA supplies the horizontal stabilizer, elevators and main gear doors.

Subassemblies are flown to Toulouse, France, in Super Guppies, which are modified Boeing Stratocruisers with turboprop engines, five of which are operated by Airbus Airlink, a subsidiary. Because each section arrives fully plumbed, wired and tested, final assembly accounts for only five percent of the total construction process. After painting and flight testing, a new airplane is flown to Hamburg, West Germany, for interior furnishings and cargo equipment before returning to Toulouse for customer acceptance.

Airbus Industries success has been remarkable. In the 20 years since its entry into the bitterly competitive airliner market long dominated by Boeing, McDonnell Douglas, and Lockheed, 1,172 of its airliners had been ordered by 78 airlines, seven of them American. That's 1,172 airliners that will not roll off assembly lines in Seattle and southern California. As recently as 10 years ago who would have dreamed it possible?

I stood beside the remarkable A320 talking with the long-time friends who within months will join me in retirement. I think they envied me my life without ground schools, simulator training, and checkrides that last all night. Strangely enough, I envied them.

Postscript: "Airbus" is not the name pilots would have chosen for an airliner. The connotation bothers them. Interestingly, the title did not originate with the European builders who use it today.

Sixty years ago an airplane christened the "Airbus" was wheeled out of the Delaware factory of Bellanca Aircraft Corp. It was as remarkable for its time as the A320, and almost as unusual. A stately semi-sesquiplane (a biplane having one wing with not more than half the area of the other) with a wingspan of 65 feet, it accommodated up to 15 passengers and had a toilet.

Bellanca's Airbus cruised at 135 mph, respectable performance for a single 650-hp radial and fixed-pitch propeller. The Army version (C-27) was used for freight hauling. Fitted with floats, the Airbus was a favorite of bush pilots. Two were still flying into wilderness lakes 35 years after the first was built.

Neither is a side stick control in large aircraft without precedent. Late model four-engine B-24 bombers had a similar control on the left side of the cockpit that directed the autopilot. Called the *formation stick*, it was intended to ease the first pilot's workload on long missions but it failed to provide the immediate subtle control adjustments required in tight formation flying. "Their heart was in the right place but it just didn't pan out," recalls one combat veteran.

41

Old Dumbo

YOU SIT WAY UP THERE in a C-46, three feet higher than in a 727. I had forgotten that, but the cockpit was vaguely familiar. It is primitive, a strictly functional workplace seemingly the result of afterthought, as though the designers almost forgot a place for the crew. The pedestal mounts crude levers better suited to a bulldozer. A wheel/cable arrangement at each side window selects fuel source. The large control wheel and oversized trim controls hint at the effort required to maneuver.

From the copilot's seat I watched the languid turning of the right prop and counted. "Six blades, switch on." The silver prop revolved twice more then, with a muffled grunt and belch of smoke, the big Pratt went to work and settled into rhythmic idling. Then Dave Cummings brought his side to life. We watched the faded needles move in the ancient gauges. Oil pressure, fuel pressure, manifold pressure, revs: The after-start drill is indelibly etched in memory.

Dave has taxiing down cold. The coordination of throttles, brakes, and rudder required in large taildraggers is almost a lost art. Today's trike driver with nosewheel steering doesn't know he's born. After runup, Dave took position, locked the tailwheel and eased forward on the throttles. We lumbered off and were quickly airborne. The runways this old C-46 was built for rarely exceeded 5,000 feet.

So this is how it was. Forty-two years of other airplanes had passed since I last strapped into that high seat as an instructor at Reno Army Air Base. It was sobering to know that if we flew this tired old plane to O'Hare or Andrews Air Force Base, most pilots would ask, "What's that?" The few who knew would likely call the C-46 a flying mistake. It never shook that reputation. Those who flew it tell another story, however.

Curtiss-Wright was a name to reckon with through World War II. It built airplanes, engines, and propellers. The 1939 *Jane's All the World's Aircraft*

Douglas C-124 was in its day one of the largest cargo planes flying.

devoted seven pages to Curtiss airplanes. Douglas, Lockheed, and North American rated four pages each, Boeing and Martin three.

The DC-3 had yet to fly when Chief Engineer George Page envisioned a Curtiss competitor. The following year plans were drawn for a twin-engine pressurized airliner with 36 seats and underfloor bins for 8,200 pounds of cargo. It would be big, with a double-lobe fuselage (inspiring the nickname, "Pregnant Whale"), hydraulically-boosted controls and a wingspan four feet greater than that of the four-engine B-17 *Flying Fortress*. The CW-20T, with twin rudders and Wright R-2600 engines, was first flown in early 1940 by Boeing test pilot Eddie Allen.

Many modifications followed, including installation of Pratt & Whitney R-2800 engines, an improved landing gear, and redesign of the tail to include a single rudder. By mid-1940, the war situation in Europe had deteriorated; American involvement appeared probable. Curtiss was told to forget plush airliners and think military transports. A stressed floor and cargo doors were stipulated; the result was designated the C-46 Commando for Army duty, the R5C for the Navy and Marines.

Despite Page's protests that his brainchild needed more work, 46 were ordered. The first version accommodated 50 troops or 33 litter patients plus medics—or 10,000 pounds of freight, a considerable increase over the capacity of the C-47, the militarized DC-3.

As predicted, bugs appeared. The control boost system proved difficult to keep in repair. The rudder boost eventually was removed and later, those for ailerons and elevator. Vapor locks at altitude were in due time eliminated by installing submerged pumps in fuel tanks. Mysterious in-flight explosions at first blamed on hazardous cargo were traced to inadequately vented wings.

And, of course, those propellers. Oh, the horror stories told about Curtiss electric props! Impulses to the pitch-changing motor in the hub were transmitted through brushes riding on a series of spinning rings. Dirt or grease on the assembly could cause loss of pitch control to the extent blades went into flat (high drag) pitch and could not be feathered, a disastrous malfunction. Regular inspection and cleaning virtually eliminated the problem, however. Veterans who repeat the wild tales forget the P-47 Thunderbolt, P-61 Black Widow and B-26 Marauder (which suffered the lowest combat losses of any American bomber)—all had Curtiss electric props.

To compare the C-46's teething troubles with the DC-3's smooth transition to military duty is hardly fair. Five years of airline experience contributed to the C-47's remarkable success. The C-46 enjoyed no such shakedown in the hands of professional pilots and mechanics. Instead, while still under development, it was assigned to overseas routes including the infamous "Hump," the supply route between India and China across the Himalayas.

In December, 1942, the Japanese captured the Burma Road, China's last link with its allies. President Roosevelt proposed to supply Chiang Kai-shek and U.S. forces by air until a new road could be built, an undertaking without precedent. The air route lay across the world's highest, most treacherous mountains, and through appalling weather. There were no en route navigation aids; during monsoons, arrivals let down through low ceilings in torrential rains to find runways under water. A clear day was a mixed blessing, for then the lumbering transports were easy targets for enemy fighters.

Transports of the day could scarcely maintain the altitudes required even in smooth air. Thunderstorms, winds of jet stream force, and ice were recurring perils that could force descent. Living conditions were miserable; malaria and dysentery were rampant. There were never enough parts for proper maintenance. The China-Burma-India (CBI) corner of the war was indeed the "forgotten theater." It was there, halfway around the world on the "Rock Pile," that the big Curtiss and its young crews wrote air transport history. A Hump tour (500 to 650 hours) earned the DFC.

The initial goal of 2,500 tons a month was raised to 4,000 then to an "impossible" 10,000. Three years later, in July, 1945, 71,042 tons crossed the Hump by air. A typical C-46 load comprised 23 55-gallon barrels of 100-octane gasoline and 3,000 pounds of bomb fuses. In all, 650,000 tons of critical supplies were airlifted from India to China. When the war ended, Air Transport Command had 330 C-46s, 167 C-47s, and 132 C-54s flying the Hump. The cost was high, more than 1,000 lives and 600 airplanes, but a million of the enemy were kept occupied by forces supplied by air.

After returning from C-47 flying in Africa, I was sent to Reno for C-46 training and retained as an instructor. The ground school followed the nuts-and-bolts approach then in vogue; I believe we could have built a C-46 upon graduation. The first dual was a shock; the big brute handled like a moving van with front-end problems. Flying required two hands, two feet and muscle from chocks to chocks. Moving pianos was as easy.

But it was stable, mushed straight ahead when stalled, did remarkably well on single engine, and was rock solid on instruments once you learned to trim. The cockpit was a mess. The sloping windshield made the runway seem to curve away in the distance. A night landing in snow or rain in a gusting crosswind was true adventure. Yet old Dumbo (for Walt's flying elephant) grew on you. It was built like a brick privy, could haul the freight, and kept plowing along with a frightening coat of ice. I speak from experience. Reno in winter was good training for the Hump. The pilot who passed the course would find everything that followed easier, including jumbos four decades later.

The Confederate Air Force keeps two C-46s airworthy. Colonels Dave Cummings, Chet Brakefield, and Sublett Scott (every CAF member is a colonel) take good care of one in Oklahoma. *Tinker Belle*, a C-46F, was one of the last of 3,182 Commandos built. All C-46Fs were equipped with hydraulic props (dubbed "toothpicks") instead of Curtiss electrics ("paddle blades"). We had a couple with toothpicks at Reno. The electric prop's four blades got you airborne quicker while the hydraulic's three afforded a slightly faster cruise.

Dave pointed to my wheel. It was set in concrete. When pulled with the effort needed to chin, the altimeter moved a hair. "Needs a grease job," I said. Time dims memory but I don't think our hydraulically-boosted controls were this stiff. I'd like to watch the expressions of jet airliner pilots invited to that cockpit. The old tub responded reluctantly as if to say, "You've grown soft with all that newfangled power steering." An airplane can cut you down to size; *Tinker Belle* didn't think much of me.

The C-46 didn't make it as a luxury airliner after the war but found work with 341 carriers, mainly as a freighter. A number of surplus transports crashed due to poor maintenance and inept handling. When a C-47 went down, pilot error was assumed; a C-46 crash revived all the "Curtiss Calamity" tales. A television documentary included the inane comment, "The C-46 was an airplane so badly designed it should never have been allowed to fly." In fact, the C-46 was, to quote a noted air historian, "one of the most reliable military aircraft of the Second World War." Indeed, it remained on active duty for 27 years, astounding longevity for a military type.

Keeping that C-46 alive is not all fun and games. Before we could fly, the three colonels sweated all day over an rpm drop, three times dismantling a magneto under the August sun. You have to admire those gray-headed enthusiasts; they don't sit around talking about the old days. They do things. In my book they reflect the best side of aviation. They could use more help, preferably from young buffs interested in preserving history. Of course, money is the big problem. Air shows reimburse fuel and crew expenses but there are other costs to keeping this worthy relic flying. You old C-46 drivers, go ask Chet and his friends to show you *Tinker Belle*. You just might catch the fever.

It was dusk when we turned onto a long final. "Gear down, quarter flaps." The mains locked into place with well-remembered thumps. Dave called for more flaps and retrimmed. We sank toward earth, those magnificent Pratts rumbling, drifted across the fence and settled onto the concrete tail low with that distant squeal of tires that marks the perfect ending of flight. Dave swung around, careful to keep the inboard wheel turning, and cut the mixtures. I sat there, remembering long-forgotten people and places, good times and bad. And listening to the sounds of big radials cooling.

42

Air travel 50 years ago

NOTHING TURNED HEADS FASTER in the early 1930s than a casually dropped, "I flew back from Chicago." Few people had been off the ground. Air travel was seen as an expensive and risky novelty for wealthy executives and movie stars, an image not entirely undeserved.

Most people by 1938 at least knew someone who had chanced it, survived, and in most cases wanted to do it again. The airlines were catching on, thanks to imaginative advertising and improving reliability and safety. World War II would soon give aviation a tremendous boost. In the meantime, during those last fair days before Pearl Harbor, a trip by air was high adventure.

The 80-page *Official Aviation Guide of the Airways* listed 23 domestic carriers. American, United, TWA, and Eastern were the "Big Four." Some 224 cities were listed as "regular or flag stops." All seats were first class, all fares identical. New York to Chicago cost $44.95 with 10 percent off for a round-trip ticket, no matter how you went. Mind you, $275 a month was good pay then.

(Actually, New York had no airport in 1938. Its flights originated across the Hudson River at Newark. LaGuardia would not open until late the following year and JFK a decade later.)

Which to take? American flew "Flagships," each named for a city or state the line served. Admiral's pennants fluttered above the cockpit when a Flagship taxied in (and woe betide the copilot who forgot to retrieve his before takeoff). You could fly nonstop from New York to Chicago on the *American Eagle* or the *American Arrow* in just 4:45, or the *New Yorker* or *Niagara* with stops in Buffalo and Detroit, the latter two promising a view of Niagara Falls.

United whisked you there in a "*Mainliner*" also named for a city or state. You could catch the nonstop *Overland Flyer* or the *Californian* with a stop in Cleveland. The line's title and trademark red, white, and blue shield subtly hinted that it was in some way "officially" endorsed. United was "The Main Line Airway."

Transcontinental & Western Air (TWA) was "The Lindbergh Line," the Lone Eagle having laid out its routes. Its airplanes bore no names but its trips did. New York-Chicago service was delivered by the *Sky Rocket, Sky King, Sun Pacer,* and *Nighthawk.* The "Skyclubs" flown by day became "Skysleepers" by night. The coast-to-coast berth surcharge was $8.

War ace Capt. Eddie Rickenbacker, Eastern's boss, disdained airplane and trip names, perhaps because he didn't think of them first. His line flew from New York and Chicago to Florida.

Ad writers pulled out all the stops. "Hot air is turbulent air. Cool air is smooth air," explained TWA. "Our pilots plan each trip so as to fly at smooth, cool levels." United claimed its Mainliners could make 227 mph and "loaf along at 205 using only 60 percent of horsepower." American boasted of its "cool Southern All-Year Route." In fact, there was not a nickel's difference between Skyclubs, Mainliners, Flagships, and Eastern's "Great Silver Fleet," all being stock DC-3s that ambled along at 175, nor was there a prayer of cool, smooth flight every time during summer months.

Your call for a reservation was answered by a person, not a recording. Tickets were picked up at the airport or downtown ticket office, which was in the most prestigious hotel. You were asked to be at the airport 15 minutes early for check in. Of course, you immediately canceled if your plans changed; "no-show" was not in air travel's lexicon.

The carpet at the terminal ticket counter concealed scales. To ask a corpulent matron her poundage was unthinkable but passenger weights were required on the manifest. All passengers were in their Sunday best. Flying was an uncommon experience; you dressed accordingly.

At last, the big moment. The flight was called and with feigned nonchalance passengers walked from the terminal toward the sparkling Douglas beyond the gate. There was no metal detector, no x-raying of handbags, no police watching. You showed your ticket to the stewardess and walked uphill to any seat you liked. The DC-3 accommodated 21 fares, 14 in seven rows on the left of the aisle, seven on the right. The seats were as comfortable and the leg room better than are found in today's typical coach cabin.

The cabin door was closed, the steps rolled away and the engines came to life. The airplane trembled expectantly; you knew exciting things were about to happen. The Three waddled off nose held proudly high and, after engine run up, began its race across the grass; few fields had hard surface runways. In seconds the uneven roll ceased and rock steady flight began. Engine noise during takeoff was not uncomfortable and climb power made possible near-normal conversation. As the earth receded, houses became symbols on a Monopoly board with cars crawling among them like ants, and then you were looking down at miniature farms and broad rivers reduced to streams.

The young airlines gave much attention to cabin decor, using subdued colors and expensive fabrics to promote an air of plush restfulness. Individual air vents and reading lights, magazines in embossed leather binders, and detailed route maps made the trip seem worth its price. Norman Rockwell painted a little old lady in a DC-3 with the map on her lap, making her first air trip and obviously having the time of her life. It is a perfect portrayal of air travel 50 years ago. A framed print of it hangs in our den, prized all the more because the famed artist signed it for me.

For its day the DC-3 was exactly right in so many ways, not the least of which was its size. Twenty-one was an ideal number of people to introduce to flight for rare was the trip without a "first rider" or two. The atmosphere was intimate, relaxed and conducive to friendly across-the-aisle chatting. There was a feeling of sharing the unusual, of being given a glimpse of the future. There was time to take in the view and a view to take in, time to read or be lulled into a nap by the engines' hypnotic drone. Most of the time it was a very nice way to travel.

Common practice included a climb to 300 feet above cruising altitude, then a shallow dive to "get her on the step." An announcement form with the blanks filled in would be handed back along the seats:

"Welcome aboard Flight 24. We are cruising at *8,000* feet. Due to *tailwinds* our groundspeed is *190* miles an hour. Our estimated time of arrival at *Dallas* is *2:45 p.m.* Please tell your stewardess, Miss *Amber Criddle* if there is anything we can do to make your trip more enjoyable.—Capt. *Brian Marks*." You already knew about Brian and Amber as the crew names were posted on the cockpit door.

Stewardess Criddle kept busy answering questions, pointing out landmarks and passing out gum and playing cards. There was complimentary coffee, tea, and soft drinks but no liquor. You didn't drink and fly, even as a passenger. She would stamp and mail your letters, and who could resist impressing friends with airline stationery?

A typical lunch included soup, crisp salad and a sandwich served on a pillow on your lap. There was also an apple, some mints and a box of five cigarettes. The far more elaborate tables set on long premium runs would please today's first-class passengers. The captain would come back to answer questions, reassure the apprehensive, and take small boys up for a look at the incredibly complex cockpit. (Such a treat aboard a DC-2 cemented this one's airline ambitions.)

All too soon it would end with a slow drift earthward, then the Fasten Seat Belts and No Smoking lights came on. The houses grew larger, rushing by beneath the wings, and you felt the soft slide of tires on clover. Tail down and proud nose raised, the Three would swing around at the gate and stop, its engines expiring in a death rattle of reduction gears. And you'd walk off, taking little notice of the curious lining the fence and wondering what it must be like to fly.

Looking back, it was all so old-fashioned, so quaint. We thought we had found the answer—but we didn't have a clue, did we?

43

JFK-LHR: 3:35

BRITISH AIRWAYS FLIGHT 9092 was gently nudged from its gate right on time. My ticket read 11:30 and that was precisely when we moved—not a minute sooner or later.

That's the way to do it. An on-time departure means the airline cares, that it takes pride in its service, and suggests the rest of the trip will be marked by the same attention to detail. Weather and traffic can unavoidably delay an arrival, but a push back ten minutes late is the trademark of slack planning. Aside from a mechanical problem or gate hold ordered by ground control, there is little excuse for not blocking out on the dot.

We slowly taxied along Kennedy's uneven taxiways, paused while a DC-10 started its roll, then took position on Runway 22R. A minute passed. I glanced about the lean 100-seat cabin. The comfortable pewter leather seats looked brand new. The carpeting could have been laid yesterday. There was none of the worn appearance usual in older airplanes. This ship might have been delivered last week. In fact it has logged hundreds of transatlantic crossings in the last two decades.

But G-BOAC is no ordinary airliner and Flight 9092 is unique travel that suggests a special effort at punctuality. G-BOAC is a Concorde. The muted turbine rumble increased to a growl and then a distant roar. Afterburners on, brakes off. The rate of acceleration was impressive and the time spent gathering speed even more so, but we had been advised we'd need 215 miles an hour to fly.

The cabin tilted sharply, the last few feet of concrete flashed past and we swept across Jamaica Bay nose high, rapidly gaining altitude. What a sight we must have been from below. At speeds less than 300 mph on the bulkhead display, there was the mild shuddering felt in conventional types with flaps extended. It's a low speed characteristic of its delta wing; the Concorde has no flaps. Once above 10,000 feet and well south of Long Island, speed rapidly built and the ride became rock solid. Our seats could have been bolted to the hangar floor.

Five years earlier my wife and I took the *Queen Elizabeth II* to England and returned on the Concorde. It was a post-retirement treat, a once-in-a-lifetime adventure for two travel aficionados. The round-trip by sea and air was a marvelous success, something to be remembered forever. Our son, Terry, watched our videotape, heard our enthusiastic account and said, "I'm envious!" I never dreamed of repeating any of it and he couldn't manage it then, being fully occupied rebuilding a flying career shattered by Braniff's collapse.

His life, like that of many other pilots since deregulation, has been dives and zooms that if plotted on a graph would resemble a roller coaster. Uncertainty goes with the job. He's a USAir captain today and a 737-200 check airman. When he said, "Let's go to London on the Concorde," I said, "Give me five minutes to pack."

The Mach number reached .82, a velocity familiar to subsonic jet pilots and maintained its steady advance to .9, .95, .99. "Incredible!" said Terry as it flicked to 1.0, 1.2, and 1.3. "I'm not believing this." We were passing through 37,000 feet in the slow climb that would continue until descent began.

"Try one of these cheese things," I said, passing the canapé tray. "They're quite good."

"How can you sit there talking about food when we're going faster that the speed of sound?"

"What? It's one point five, is it? Good show. Now we'll really get cracking," I said, mimicking my RAF cohorts of long ago.

My casualness was pretended. The second time around was every bit the fun of the first. On the Concorde you feel a part of something extraordinary, of experiencing flight as it was meant to be—and of course you are. We should have much more supersonic air travel.

One of my DC-7 skippers was incensed by the circuitous routings and reduced speeds enforced at busy terminals. "Damn it!" he'd howl as we entered a holding pattern with a 45-minute EAC time. "An airplane is supposed to go direct from A to B at full speed!" By the same token, why are we plodding along at 600 mph an hour when the British and French proved 20 years ago that 1,350 is feasible?

The Russian Tu-144, the first supersonic airliner, first flew on New Year's Eve, 1968, the Concorde 60 days later. Our entry was to have flown in 1972 and what a magnificent aircraft it would have been.

Characteristically, Boeing thought big. Its 317-ton 2707-300 was to be 62 feet longer than the 747, cruise at Mach 2.7, and seat from 250 to 320 passengers. Projected seat-mile operating expenses promised reasonable fares for long international trips.

President Nixon gave the go-ahead for construction of two prototypes. The federal backing required was to be repaid upon delivery of the 300th aircraft. If as many as 500 were built, the government would realize a $1 billion return on its investment. With 122 orders in hand from 26 airlines the designers went to work. Then, shamefully, the Senate voted not to fund the project.

The Tu-144 proved unsuccessful and was abandoned. We refused to try. The Concorde is a galling reminder of what we should have done. Britain and France are justifiably proud of it. If their marvel had received the welcome accorded American airliners in Europe since before World War II, more carriers would have flown it. In fact, 16 airlines placed 74 orders for Concordes but landing rights at New York and Washington were bitterly opposed. Had the petulant busybodies who protested been around in 1903, Orville and Wilbur would have been evicted from Kitty Hawk for disturbing the seagulls.

Oceans cover three quarters of the globe. Supersonic passenger service across them is practical and inevitable. The question is not if there will be more SSTs, but who will build them.

For lunch I had the "Maine lobster with king prawls, cooked in a mild curry-flavoured sauce and garnished with basmati rice" while Terry chose the "Fillet of beef seared on a hot griddle and seasoned with pistachio butter," both complemented by exceptional vintages.

Captain Mike Riley invited us up front. He's a Concorde instructor whose unpublished manuscript on Concorde piloting is fascinating. We were creeping though 56,000 feet with autopilots slaved to Mach 2.0 and altitude slowly increasing as we grew lighter. "On this run we'll burn about the same fuel as a 747," he said. It is surprisingly quiet in the cockpit; there is little air 10 miles up to create wind noise. On the other hand, skin friction lengthens the fuselage 10 inches and leaves cabin walls warm to the touch.

First Officer Alan Atkinson penciled our 22.5 miles-a-minute progress on his chart as casually as I used to note three miles accomplished by a DC-3 in the same time. As a boy my father was thrilled by mile-a-minute trains yet he lived to travel ten times as fast in a 747. I remember airliners that did two miles a minute and lived to see Mach 2.0 passenger service. What will my grandchildren see during the next half century? The tantalizing challenge is to do still better and it will not be ignored. Some of the solutions should be stamped "Made-in-USA."

Second Officer Phil Newman lives with his wife in Southern California and commutes to London. He's flown 4,000 supersonic hours in 12 years. His complex engineer's panel makes it easy to believe Concorde crew training requires six months. Our ship, G-BOAC, has logged 13,000 hours and completed more than 5,000 "cycles"—takeoff and landings. British Airways' seven Concordes have racked up a total of 85,000 hours and 31,000 cycles, yet supersonic air travel remains a novelty. The challenge is to make it economical.

All too soon it was over and we began descent. Speed rapidly dwindled to familiar subsonic Mach numbers and then with gear extended to the slow velocities that pleased us in piston days. We banked around the famous courts at Wimbledon, rode smoothly down final at 150 knots, and settled ever so lightly onto 28L at Heathrow.

I made a note: "JFK-LHR: 3:35" and thought again, "Now *that's* the way to do it."

44

Mach 25? You bet!

"HEAVIER-THAN-AIR FLYING MACHINES ARE IMPOSSIBLE," said Lord Kelvin, president of Britain's prestigious Royal Society in 1895. Charles Duell, director of the U.S. Patent Office, went further. "Everything than can be invented has been invented," he said in 1899. Four years later the Wrights proved them both wrong. At first the brothers themselves foresaw no commercial application for their invention, yet Orville lived to ride in a Constellation.

I poke no fun. At the turn of the century there was little recent scientific progress on which to base educated guesses about the future. It soon became clear that long-range predictions about flying were risky, no matter how optimistic, and a pessimistic forecast was downright foolish. The more often a new concept was dismissed as unrealistic, the more likely someone made it work.

More difficult to understand was the respected engineer who wrote, "500 m.p.h. Can't Be Done!" in 1930. His article appeared in the forerunner to *FLYING*, *Popular Aviation*, which cost 25 cents. With diagrams and formulas he conclusively "proved" that 400 was about the fastest we'd ever fly. His guess was close, considering the state of the art at that time. Few propeller-driven airplanes with reciprocating engines have sustained more than 450. He did not dream of turbine and rocket power or pressurized flight in and beyond the atmosphere's outer fringes.

I reread his piece aboard a return flight from London. He scoffed at "air-dreamers" who talked of flying at 500 to 1,000 miles an hour. Considering the astounding progress made since the Kitty Hawk success 27 years earlier, his dour prediction is somewhat surprising. You'd think the possibilities of revolutionary new powerplants and pressurization would have tempered his forecast. I wondered if he lived another 17 years to see Yeager push the X-1 to 670 miles an hour or until 1954 when he reached Mach 2.4. It was time for lunch. I put the magazine away, noting that our Concorde was then 11 miles high and making twice the speed of sound.

A week earlier I stood on the stone pier at Plymouth where the U.S. Navy's NC-4 crew stepped ashore in 1919 after the first transatlantic flight. In a London museum I admired the Vickers Vimy in which Alcock and Brown made the first nonstop crossing a month later. I thought of Hawker and Grieve, Nungesser and Coli, and the other brave souls who perished trying to cross the Atlantic by air. An awareness of air history gives special meaning to a Concorde trip and perhaps a glimpse of the future.

I visited the Marine Terminal at newly-opened LaGuardia Field in 1940. A Pan American Boeing 314 lay at anchor beside an Imperial Airways Short flying boat. Parked among the DC-2s and -3s was a sparkling TWA Stratoliner, the first pressurized airliner. The boast was that LaGuardia was large enough to accommodate New York's air traffic for the next 30 years.

I remembered my own transocean hours, the first crossing to bring a war weary C-53 home from Africa, then to ferry a new B-17 overseas. Years later it was 707s across the Pacific and 747s to Europe, almost casually crossing the treacherous wastes spanned at awful risk by those who did it first.

One afternoon over California in 1944 a strange silver airplane eased up alongside our lumbering transport. It had no propeller and trailed black smoke. It was a Bell P-59, the first "jet" any of us had seen. It was a hint of the future yet the wildest prediction we could have made wouldn't come close to what's happened since. The sound of reciprocating engines has long since become a memory at airline terminals.

Long-time friend, John Taylor, then editor of *Jane's All the World's Aircraft*, said, "Concorde is what we worked for all these years. It's everything." And it is, for now at least. By the time ours landed at Kennedy I had much the same feeling as when, 18 years earlier, I walked through the wooden mockup of the 747; the prototype had yet to fly. The Concorde and 747 have shown what can be done with the speed and size of commercial aircraft. Each is logical and inevitable. Yet, as marvelous as they seem, they are but rungs in the ladder. It is less what they are than what they promise.

Critics said the jumbo was too big and would bankrupt its owners, repeating their appraisal of the 707 15 years earlier. As of August, 1987, 809 had been ordered. They shrugged off the SST as a brilliant freak without a prayer of competing in the air travel market. Yet, 10 years after entering service it was earning profits. Our Concorde had logged 10,000 hours and exceeded the speed of sound more than 2,700 times. Its crew of seven had 50 years of supersonic experience.

The 747 and Concorde opened doors that will never be closed. History proves the risk of predicting the future of aviation, at least, of declaring what cannot be done. Visionaries who did not understand the problems have been proved right more often than technicians who did.

President Reagan in 1986 endorsed research on an air/spacecraft able to cruise-orbit from Washington to Tokyo in two hours at 12 times the Concorde's speed. The engineering problems to be solved are staggering, not to mention the funding required. There is no question that such a radical craft will be built, in fact, an experimental "National Aero-Space Plane" (NASP) dubbed the X-30 could be flying/orbiting before the end of the millennium.

NASP, a joint NASA-Department of Defense design project, is expected to operate from existing runways, fly in the upper atmosphere at Mach 6 to Mach 12 and orbit at Mach 25. The prototype could be operational within 12 years. Its commercial possibilities are tantalizing. Presumably, the passenger-carrying "Orient Express" would scramble at an unbelievable rate to an altitude measured in miles from which its shock waves would not pose a problem.

Meanwhile, Boeing and McDonnell Douglas each say they can build an interim Concorde replacement able to carry 250 to 300 passengers at Mach 3 to Mach 5 nonstop over long transpacific routes. Airline analysts say Pacific passenger loads will exceed transatlantic traffic by 25 percent in the year 2000. Interestingly, the sonic boom problem will be minimal because such routes as Los Angeles-Sydney and San Francisco-Hong Kong can easily be flight planned to avoid crossing inhabited land.

"We can build an aircraft to meet those requirements," said a Boeing official, "but can we do it for what the airlines can afford to pay?" Studies by McDonnell Douglas suggest that 9,600 airliners will be required by the year 2000, while by 2025 some 1,250 could be supersonics. Neither builder is likely to tackle an SST project without government assistance and/or the involvement of foreign associates. If we are to capture a large share of the estimated $200 billion high-speed market we must be a major partner in any multinational consortium. A NASA spokesman said, "This market will be an important factor in the U.S. sustaining a positive balance of trade in aerospace products."

The Alcock and Brown Vimy, a converted World War I biplane bomber, weighed 12,500 pounds loaded, climbed to 5,000 feet in 22 minutes, and struggled along at 90 mph. Its success inspired predictions that regular passenger service between the Old World and the New was just around the corner. International airlines were organized and stock was sold. Plans were drawn for huge multiengine aircraft with 10 to 30 seats. Insiders knew better; they knew such ideas were premature. Some said it would never happen. But, except for their timing, the dreamers were right.

In due time the Concorde and 747 will take their places alongside the ancient Vimy, museum pieces equally curious and outdated. People will walk through them, peer into their cluttered cockpits and shake their heads. Who says there won't be two-hour airline service from here to Japan?

Not I.

45

Capacity woes

AIRPORT CONGESTION IS NOTHING NEW. When I watched flying at our field, congestion was two DC-3s loading and a third on final approach. Saturation was three on the ramp because there was no room for more; there was only one gas truck. Inside the small WPA-built terminal were two ticket counters and a waiting room. It was rarely filled; there were only 10 departures a day. Such was Kentucky's stake in air travel in 1939.

The high point of a trip to the 1940 New York World's Fair, my graduation present, was an afternoon at brand-new LaGuardia Field. Eight of the 12 gates were occupied, one by a huge TWA Stratoliner. A Pan Am flying boat lay anchored at the nearby Marine Air Terminal. LaGuardia was an exception, a glimpse of the future. A critic said, "We are using 1926 airports for 1940 planes," an accurate description of most municipal fields.

When World War II began, US domestic carriers flew 322 airplanes, 80 percent of them DC-3s. Air travel was a novelty. Fourteen 747s can seat as many fares as that prewar fleet. Walking through the crowded halls at O'Hare or Los Angeles today, I think of the travel revolution brought about by aviation in just 50 years. Even having witnessed most of it as an insider, I can hardly believe my eyes.

After the war, rapid increases in aircraft size and performance caused major airport problems. Facilities built for DC-3s were a worry for DC-7 crews and perilously marginal for jets. Kansas City Municipal was an example. A mile from downtown, it was handy for passengers, downright hairy for pilots. Only the north-south runway was usable by anything larger than a Convair. A circling approach at night in rain or snow was of special interest in a DC-3, exciting in a DC-7, hair-raising in a 707. "You don't land a 707 here, you gamble with it," grumbled a skipper emerging from the cloud of slush thrown up by hard reversing.

Extending runways (impossible at KC, of course) was a temporary fix, strength being as critical as length. Runways suitable for the Stratocruiser, the

Gear up! Braniff 727 departing Love Field at Dallas.

largest piston airliner, cracked beneath the first generation of jets and needed complete rebuilding to accommodate jumbos the weight of five Stratocruisers.

Improvements came much too slowly for pilots. They watched in exasperation as funds that should have gone into runways and navigation aids were often frittered away on ornate terminals with moving sidewalks and fancy restaurants. At such "monuments to mayors" they sweated out maximum weight landings on crumbling concrete made treacherously slick by tire residue.

Aircraft performance and weight stabilized with the 747's arrival. The typical airliner of the foreseeable future will be subsonic; the largest will weigh about 450 tons. The major problem since has been not been runways, but the number of airplanes they can handle. Conga lines of departures were a familiar winter sight at Chicago and New York by 1969 when TIME commented that "airports are straining at the seams." Today your New York-bound trip might sit at its gate in Kansas City or Atlanta for an hour past scheduled departure time awaiting airspace in which to fly. An average month sees 80 percent of airline trips operating on time, officially defined as operating within 15 minutes of the timetable.

Doubling the runways at major cities, even if possible, would not double capacity. *The Wall Street Journal* was only half right more than 10 years ago in commenting that, "Airports aren't keeping up with passenger growth. They are trying to put a quart of air travel into a pint bottle." The present airway system simply cannot cope with more traffic than it now services during rush hours. Ground and en route delays along with time lost during hub transfers can reduce point-to-point times of Mach .8 jets to those attained by DC-3s half a century ago.

More airports are needed, but where to put them and how to fund them? Proposals for new fields are shouted down for environmental and economic reasons. Every city wants good air service but not the accompanying inconvenience and expense. Huge DFW International opened in the early 1970s; Denver alone is following suit. Other cities will have to make do with what they have, further improving existing facilities.

American major and national carriers operated 3,964 aircraft in 1988, not including 1,801 regional airliners. The total fleet of operators—scheduled, supplemental, commuter, air taxi, and cargo carriers—doing business at city airports numbered 5,022.

Boeing, McDonnell Douglas and Airbus Industrie had 3,100 airliners in their (firm) order books at one time, about half of which were destined to go to U.S. buyers. If Delta, American, United, and Northwest exercise their options, they will absorb 1,784 new airplanes during the 1990s.

A recent release: Federal Express has ordered 50 Cessna Caravans, with options on 100 more, to enter service late this year following—get this—completion of existing orders for 200! Fed Ex is further enlarging its fleet with 58 737s and 727s formerly operated by United and Air Canada.

Airline pilots of my generation find such numbers amazing. We remember when an order for 10 new airplanes was stupendous news. That translated into promotion and job security. We were confident the industry would grow but never imagined the air travel boom since deregulation. For every passenger we carried 30 years ago, 17 are flying today.

Even more astounding is the calm prediction that the billion passengers now flying worldwide annually will increase to four billion in the next half century. That's 77 percent of the 1988 global population, a prospect to stagger the imagination.

Perhaps the most remarkable thing about all this is the lack of serious response to the capacity problem. Where are all the new airplanes to land? Seventeen of our terminals are by definition "congested," meaning they cannot accept more traffic; the number will increase to 58 by the end of the century unless new airports are built and ways found to dramatically increase traffic flow. Lengthening a runway here and adding one there won't get the job done, nor can the present air traffic control system begin to meet future demands.

Why are we failing to take decisive action? Robert Crandall, American Airlines' president said, "It's because Congress and the executive branch do not consider aviation a top priority. We lack a sense of national purpose because we do not have a national aviation policy. There is not now—nor has there ever been—a clear-cut statement that says to everyone, 'This is our national policy on airports, on airways, on everything to do with aviation.' We ought to have one. We have to have one."

Not everyone is waiting for Washington to act. After decades of bickering, Dallas and Ft. Worth joined forces to build a huge airport midway between them. Opened in 1974 with three runways, it now has six with two more projected. You can fly nonstop to 15 countries and 115 domestic points. With 103 gates and 1,900 aircraft movements a day, at one point DFW ranked second in the nation in traffic handled. It is reported to have a $6.2 billion annual impact on the North Texas economy. It was no surprise that American Airlines relocated its headquarters there.

Houston, with a population 300,000 greater than Dallas and Ft. Worth combined, sat by and watched. Houston Intercontinental at the time ranked 32nd in traffic.

None of which was lost on Denver's city fathers. All operations at Stapleton International will be moved in 1993 to a new location 18 miles northeast of the city. The dirt is flying on a 53-square-mile tract large enough to contain DFW and O'Hare with room to spare. The Mile High City, presently the fifth busiest air travel center, does not intend to lose its edge.

Business is steadily gravitating to cities with good air service. Those that provide it will enjoy the rewards. The rest will be way stops on the vast global network now taking form. In the short term, lack of capacity might become a crisis of epic proportions. But in due time all the problems will be solved.

Looking back on that little grass field with three gates, and on how far we've come since, how could I think otherwise?

46

Love Field

LOVE FIELD CELEBRATED ITS 70TH BIRTHDAY November 23, 1987. Anyone like me who flew from good old Love for 25 years has a fondness for the place. It was part of our lives. When I was hired there after the war, our line employed 1,100 people, 210 of them pilots. There was a personal involvement that long since was erased by expansion. Today you're a number.

Back then the airline was our family, Douglases were the tools of the trade, and Love Field was home. It was "our" field. Everyone knew Kansas City belonged to TWA, Denver to United, Miami to Eastern, and all the oceans to Pan Am. Dallas was ours.

Such territorial claims annoyed the competition. A vexed American Airlines skipper, suspicious that we were receiving preference in traffic, would address the Love controller as, "Braniff Tower," just as our hard-noses called "United Tower" when approaching Denver.

A cotton patch north of Dallas was offered to the Army Air Service in 1917. It was named to honor Lt. Moss Lee Love, a pilot killed in San Diego. Old pictures show Curtiss JN-4 trainers; whether any Love Field graduates influenced the outcome in France is unknown, but they caught the attention of local promoters. A corporation was formed to develop the field. Barnstormers and air shows barely prevented its bankruptcy, however.

The owners sought a buyer in 1927. First, Dallas invited nearby Ft. Worth to cosponsor a joint venture but "Cowtown," bent on establishing its Meacham Field as the major Texas terminal, refused. Dallas bought the 600 acres and so began "Battle of Midway Airport" which ended 27 years later when the rivals (by order of the CAB) built DFW Airport halfway between them.

A fire station stands where the first fares stepped from Ford and Stinson Tri-Motors, Vegas, and Condors. A handsome new terminal was completed in 1940 by which time Dallas, then the nation's 31st largest city, ranked seventh in passengers. A year later the war began and the Army came back.

Braniff's "Big Orange" lifting from Dallas for run to Hawaii.

On my first trip to Love, I delivered a C-53 to the 5th Ferrying Group; they would see it to Alaska. The 5th occupied wooden structures on the west side of the field. Its ramp was filled with fighters, bombers, and transports awaiting ferrying crews. In a large wooden hangar, P-38s were modified. The ride into town was through open country.

After the war, Braniff leased the big hangar and moved operations, payroll, and weather office into abandoned barracks. Ground school was in the hangar, Link trainers in a Quonset hut. From this makeshift base we operated 37 DC-3s, -4s, and -6s north to Chicago, east to Memphis, west to Denver, and south to Buenos Aires.

Roofed wooden walkways affording meager protection from the weather had been extended from the terminal. The senior captain with an "airport car," a $200 junker but very much a status symbol, could park steps from his gate. The lobby was filled with passengers dressed to the nines. In those days you wore a suit for air travel, your better suit if you had two. I saw one fellow in rumpled slacks and sneakers who got away with being different. Howard Hughes.

Love had a second-story observation deck. My father loved to sit there on summer evenings, watching the airliners come and go. An inveterate ship and rail enthusiast, he became fascinated by flying.

But the airlines didn't care for anything suggesting that air travel was novel, much less exciting. Later terminals provided little hint of the action outside; the whole wonderful experience was soon desexed. One result: Now they wonder who will crew all the lovely new jets on order. I still go out to the airport just to watch, a half century after my first airplane ride.

From an operational standpoint Love left something to be desired. It was adequate in the early piston era if you paid strict attention. A departure from Runway 13 led straight across the mushrooming city; Runway 31 terminated in a sharp drop-off into Bachman Lake. The shorter north-south runway put you just above rooftops as the gear came up. I was creeping along Lemmon Avenue in a summer downpour when up ahead a DC-4 waddled across the street and broke its back in a ditch. There we no casualties but it was a pointed reminder not to come in high and fast.

A C-46 freighter settled during approach to 13, its wheels hit the bluff and it nosed over onto its back. Amazingly, its pilots survived. A DC-6 landing at dawn crashed; there were 18 survivors of the 45 aboard. The Republic F-84B was reputed to be a ground lover so I pulled over to watch when one took position on short Runway 18. With his foot in the carb, this lad released the brakes. He cleared the fence nose high with the gear coming up and plunged into a tree—or rather, through it, for he emerged still climbing. Republic built tough airplanes. The owner of a surplus P-51 ran his tanks dry on final approach, slid to a stop between houses and walked away. Its control stick is mounted in our den.

Considering its limitations and the volume and mix of traffic, Love Field enjoyed an incredible safety record throughout its long history as Dallas' city airport. The FBOs drew a steady stream of private and business aircraft; military types were frequent visitors. Eventually both runways were lengthened and a third added but by then we flew faster equipment requiring longer takeoff and landing rolls. It was still disastrous to overshoot. Landing a DC-4 without reversing during a January ice storm elevated the pulse rate. Departing in a heavy DC-7 on a blistering windless August afternoon was as thrilling, yet we were to fly 707s and 747s from the same marginal runways.

The present terminal opened in 1958 and the old one was razed. You can still drive along its entranceway to the fence, a fine place to watch the planes. The city swept out like a tidal wave, encircled the field and extended miles beyond. We built a large hangar with offices, classrooms, and a flight simulator department. The old Lockheed hangar came down and the Army base vanished. American and Delta built hangars. A 5th Ferrying Group pilot returning in 1960 wouldn't have known the place.

An 8,800-foot runway was completed in time for our Texas-Hawaii service. Flight 501's 707 used enough of it on a 100-degree to draw a sizable gallery. The first departure of our Hawaii-bound 747 went well. The next day the new runway was closed and 501 rolled down old Runway 31R, 200 pounds under maximum takeoff weight. You couldn't scrub the specifics of that takeoff from my recollection with a steel wire brush. I was the copilot.

All carriers except Southwest Airlines moved to huge DFW Airport in early 1974. At dawn four days later, DFW fell below minimums and 27 flights diverted to Love including inbound Flight 502 from Hawaii. It was as if the old place was trying to lure us back.

Recently I went back to look around. The terminal has a new silver facade to cover its original ghastly green. A multilevel garage has been completed outside the front doors. Inside, hundreds of Southwest passengers hurried to their gates. It's all different now but enough of the old appearance remains to stir up memories of things like B-36s from Carswell making practice ILS approaches on summer afternoons, then rumbling down the length of Runway 13 at 200 feet.

Down there at Gate 11 I passed time with an outbound crew that never returned. That is also where we lined up to honor retirees; you got the close-in gate on your last trip. I was down there one afternoon when no one moved or spoke. We were watching Air Force One moving out with two presidents on board, one in a casket.

Love Field witnessed a million happy reunions and fond farewells. It was a place for laughter and tears. "There has been a lot of life lived here," said a graying captain. That said it for a lot of us.

47

The view from above

WE TRIED THE COUNTRY after 27 years of city living. All those pilots with rural homes must have known something. After a long search we found three acres on top of a hill and there we built. Having just acquired a Skyhawk, I flew over to see what we looked like from the air. Only then did I realize how well we had chosen. There was the drive winding down through the trees to the house, the broad lawn and pond out back. Sites like that are rare in north Texas. We painted a 30-foot RAF roundel on the blacktop ramp outside the garage so flying friends could find us. It took six gallons of paint and two days of work.

In North Carolina, we bought a farm—well, we called it a farm. In fact, it was 13 acres with a large barn, a creek, and lots of hardwood trees. Again we were lucky. Looking down from the Cessna at the restored 120-year-old house, the tall walnuts and red barn, all surrounded by a white fence, I felt very much the country gentleman. There was nothing to match it within miles.

But it was too remote and eventually we put it up for sale. We would show prospects aerial pictures then fly them over for that spectacular bird's-eye view. How could they resist? But none of that was needed. A young couple strolled around the place and said, "We'll take it." We gave them the pictures.

The locals raved about a nearby lake so we drove over for a look. An agent showed us a number of nice houses and condos surrounded by trees. What lay beyond the trees, we wondered, and what was the percentage of permanent residents? The agent's answers were suspiciously vague so we flew over one Saturday in January. No matter how many miles of shoreline it boasted, the lake was much smaller than it appeared from the ground. The fishing might be good as they claimed but other boating possibilities were not promising.

Just beyond the trees from an area we admired was a run down farm with a pasture full of junked cars. The best section was on the far side where we saw an air strip, small country club and many attractive homes. The agent had shown us none of that, perhaps because she had no listings there.

Interestingly, the club parking lot and residential driveways were almost deserted, evidence that it was primarily a resort development, deserted in winter, crowded in summer. Forget the lake. We should have flown the reconnaissance mission first.

After a career of professional flying, the Cessna reminded me that the view from a slow airplane at low level can be useful as well as interesting. A salesman of rural and commercial property could work wonders with an airplane, quickly previewing new listings, taking pictures, and showing clients more in two hours than he could by car in two days. Considering the fees at stake ($15,000 on a $250,000 sale), flying expense would be minimal. The novelty of aerial inspection should be worth something. Has a real estate agent ever offered to show you property from the air?

Shown the advantages of aerial inspection, the average agent won't enthuse to the point of buying an airplane and learning to fly, nor should he attempt to be tour guide and pilot. Circling a ranch or factory site at low level calls for full-time attention to flying. A power transmission line crossed the fields 1,500 feet from our "farm"; I did not attempt picture taking on my own. The potential here is for enterprising owners of private aircraft and young pilots building time. (Insurance carriers should be consulted prior to any such profit-making use of a private aircraft.)

One reason older pilots and passengers have fond recollections of the DC-3 days is because of what they could see. On smooth days we stayed low, flying VFR with minor diversions to inspect points of interest: a dam under construction, barges on a river, a famous battleground, an abandoned military airfield. On the western plains can still be seen the pioneers' wagon tracks and the outlines of forgotten forts, if you're below 5,000 feet. Ghost towns and played-out mines can be spotted in the mountains if you know where to look.

The novelty of travel across the earth soon wears off for pilots. In the beginning the stern business of mastering the art precluded sightseeing and then we were absorbed with the problems of flight and ignored the landscape below. The mission, the route, the weather, and management of the machine required our attention. We came to regard flying primarily as the best way to get there and soon forgot the visual impact of our first times aloft. The cockpit is a workplace, not a seat in the bleachers.

Much of the enjoyment the Cessna afforded was in revealing at close range the sights I missed during a career of professional piloting. I was able to look down on farms and towns, trains and cars, rivers and hills as if I'd never seen them this way before—with someone else minding the store and watching for traffic, to be sure. Surprisingly, it has come to this for me: I'd almost rather sightsee than control the flight. If only I could have taken that neat little airplane to all those fabulous places I crossed at ever increasing velocity, way up there where the earth seemed drab and featureless.

In the beginning, aeronauts charged for the chance to watch them ascend. There was also profit in hopping passengers. One of the most incredible flying machines ever built was Henri Giffard's monster balloon flown at the French

International Exposition in 1878. Rising with 50 fares aboard, it was restrained from exceeding 1,600 feet by six heavy ropes, then hauled back to its pad by a steam winch. In four months, 35,000 brave souls gazed down on Paris. Clearly, people would pay to see the earth from the air.

With the advent of controlled flight, history repeated itself. Huge crowds bought tickets to watch pioneer airmen struggle around the pattern in primitive machines. Later, the more courageous onlookers bought short hops. Many the professional airman of later years who caught the fever during 10 deafening minutes in a barnstormer's Jenny. From the standpoint of insiders, those were lean and frustrating years. They wanted to see the airplane employed in serious work. Eventually it was, and they escaped the carnival atmosphere of wing walking and passenger hops.

General acceptance of air travel and the use of aircraft for business came slowly at first. A review of airline and manufacturer advertising of the Thirties and Forties brings a smile today. Flying was portrayed as swift, smooth, reliable and, while ad writers never came right out with it—safe. Too often it was none of these. "Go by air if you've time to spare," was one winter joke. The airplane's image as an exciting but dangerous novelty had to be changed; there was no hint that flying was also fun and afforded breathtaking views.

Such promotion, along with the impetus of World War II and increasing flying safety, brought about the air age we now live in. Millions ride the airlines today. Thousands of aircraft are productive business machines. And yet there are still millions who have never flown or have any inkling of the variety of work airplanes can do.

While making aviation acceptable, we desexed it. We worked hard to make air travel as commonplace as surface travel and succeeded beyond anything dreamed of 30 years ago. An airline window seat offers a poor view of what's down below and most passengers don't even get that. A poor job has been done of selling the airplane as a business tool. The real fun of flying and the almost limitless possibilities it offers beyond fast travel are known to relatively few of us, and we keep them to ourselves. We commiserate with one another about what is wrong with modern aviation. Truly, we are a smug cliquish lot with tunnel vision.

Our 172 padded along, reminding me of long ago when simply flying was all that mattered. If we pilots spent more time recalling our first impressions and thinking about all the people who have never flown, we might come up with some new ways to use what we've learned.

48

The abused privilege

AN EXPENSIVE SINGLE-ENGINE AIRPLANE with retractable gear was based at one field where we kept our airplane. I didn't know its owner or care to meet him. He's probably a businessman, but that's a guess. Whatever his purpose in flying, everyone within miles knew when he was airborne.

He'd take off at sunrise and make a high speed run across our neighborhood at 200 feet or so, presumably to let the home folks know he was up and away. Of course, he woke up a lot of other people who couldn't care less about his travels. His return was announced in similar fashion. You could copy his number without binoculars.

This grandstanding was inconsiderate and somewhat dangerous. The airport was close by and his low passes crossed the approaches to two runways. A slight miscalculation could have put him into the trees and perhaps his house—or mine. Pilots who have learned the art—and it is a special skill—of low level flight in the military or ag schools don't practice it over congested areas near airports. Our local ace was a self-taught Blue Angel.

"If he keeps that up he's going to get himself into trouble," I said to an instructor of 40 years experience.

"If he keeps that up he's going to get us all into trouble," was his dour response.

That summed it up. It's unfair, of course, but all of us are judged by the silly antics of a few self-centered boors. When one makes the news, the safe and sane hours logged by a thousand responsible pilots count for nothing in the public view. Private airmen are selfish "fat cats," said a newspaper columnist with millions of readers, "Playboys and girls with more money than brains who clutter the skies and threaten every airliner in range." The columnist said that she reached this conclusion after talking with an airline official.

My suggestion to her was that next time she talk with airline *pilots*. Many of them own airplanes; all of them know there's room in the system for everyone who plays by the rules and uses his head. She must have received a lot of letters in that vein but she didn't publish a single rebuttal.

It is a popular misconception that airline, military, and corporate pilots are pros who can be counted upon to do things right, but private pilots are mostly up there for fun and games. This image was earned in two ways:

First through deplorable public relations. What surprised me after 40 years of military and airline flying is the cliquish fraternity aura of private aviation. There's little about it to cultivate an outsider's interest. If anything, this members-only atmosphere alienates him.

The failure of this broad segment of aviation to explain and sell itself is hard to understand. There's far more to it than weekend joyriding, but you'd never know it from reading the papers and general interest magazines. In recent years private pilots have received much attention after wandering unannounced into restricted terminal areas, near-misses, collisions with airliners or each other, and far too many crashes, some of them involving people on the ground, most of them preventable.

During this time, how many features were there about the productive uses of small civil aircraft? Did you see even one about the tens of thousands of responsible private pilots who never make the front page? As an ex-airline pilot, I find this puzzling. Our end of flying didn't become the primary means of public transportation while we kept quiet about what we had to sell. Had we waited for passengers to discover air travel, most might still be riding Pullmans and ocean liners.

Secondly, the unsavory image that rankles all good private pilots is not entirely undeserved. Their ranks include too many of the wrong type. One in 10,000 is too many. These characters are incompetent, over confident, or irresponsible. Or they think that their skills in difficult ground work automatically qualifies them for difficult flying. Or they log too few hours between bottle and throttle.

There are other reasons why certain individuals should not be licensed to fly and why others should be grounded until they clean up their acts. But the odd clown with lethal flaws slips in and slips by—until he makes the big mistake and erases himself, often taking innocents with him, and private flying gets another black eye. He wouldn't slip by for long in the military; he'd be shaped up or shipped out. Airline and corporate chief pilots are equally as stern. Pros get few second warnings.

Private flying at reasonable expense is enjoyed in few countries beyond the United States and Canada. Here almost anyone of average means can learn to fly. With modest additional training he can earn the privilege of carrying passengers and making cross-country flights. For the price of a new car he can buy a good used airplane. If it is properly equipped, he can fly virtually anywhere and into any civil airport in either country. From this point on he's pretty well on his own.

His flying future largely hinges on two factors: wallet and willpower. If he's got the bucks, there's no limit to the mouth-watering toys he can acquire, up to and including jet fighters. If he's got willpower to match, he will advance deliberately, spending whatever time and effort are needed to ensure competence. A mature individual's willpower is tempered by won't power—the honest acknowledgment of personal limits and the self-discipline to remain within them.

No one is going to make him do it correctly. He can check out in new types with a fraction of the book work and dual instruction required of professionals and remain "qualified," at least on paper, without in-flight monitoring, recurrent training, or rechecks for one or two years. The degree of proficiency he attains depends upon the study, practice, and training he personally instigates.

A friend of 14,600 hours experience, 10,000 of them in the 727 (two years of that as a flight instructor), lost his job due to a bankruptcy. Starting at the bottom with another line, he flew as engineer then copilot on the 727 for four years and just got a captain's bid. Four days of ground school and simulator training would requalify him for the left seat, but that's not the airline way. Instead, he will spend three weeks in class, 10 hours in the simulator and get a 25-hour line check. During his first 100 hours on the line higher landing minimums will apply. Call it overkill if you like, but his company enjoys one of the finest safety records in airline history.

The almost unbounded freedom enjoyed by our private airmen is, in a sense, a weakness of the system. The overwhelming majority of people attracted to small airplane piloting for pleasure or business are sensible and responsible individuals. They work hard at flying and it works well for them. But they are repeatedly penalized for the foolishness of a few lethal misfits who regard flying as a right rather than a privilege and there's little done to weed them out.

For that matter, most of us accept this rare privilege too casually. My friend, Naveed "Bobby" Riaz knows its true value. He's a businessman from Lahore, Pakistan, a city somewhat larger than Chicago. In his country the restrictions are so formidable he came over here years ago to learn to fly; he soloed in a week. He comes back once a year to talk to American pilots and log a few more hours. Bobby has accumulated 140 hours this way.

Does he read *FLYING* magazine? "If and when it reaches me," he said. "Six copies of it are worth as much as our postmen make a month. Letters come through, but not always magazines." We'll meet again the next time he gets to fly, halfway around the world from his home.

At the airport where we talked I overheard one pilot telling another about a fellow who routinely flies his twin in weather without an instrument rating. He turns altitude reporting off and the autopilot on. They both laughed; they thought that was very funny.

49

Details, details

ONCE MY COLUMN INCLUDED a World War II story that began, "A B-24 in Alaska encountered a Japanese 'Betty,' a four-engine seaplane . . . ," which was wrong, of course. The "Betty" was a twin-engine bomber; the seaplane involved was an "Emily." I knew better and so did a lot of readers—so many that the same "You're-right-I-goofed" reply went out as complaints came in.

Many of *FLYING*'s readers are pilots, a breed irritated by things that aren't quite right; most of the rest are enthusiasts who keep up with current events and know some history. "United Air Lines" or "Federal Air Regulations" or "Lockheed 12A Electra" in a journal supposedly staffed by knowledgeable writers raise eyebrows as quickly as a loose fuel cap. An aviation writer's opinion is arguable but he should get his facts right. The slapdash approach to aviation topics seen in the papers and on TV won't do.

My airline was chartered as Braniff Airways, Inc., in 1930 and such was its corporate name for 52 years. Yet to Dallas reporters and newscasters we were always "Braniff Airlines." Dallas was our headquarters for Pete's sake! Do I nitpick? If you're Joe, do you like being called Jim?

As every pilot learns, carelessness about little things can lead to big problems. A BAC 1-11 crew with the laudable intent of making up time hurried through the checklist and took off, only to discover they had not been refueled. Miserable with embarrassment, they returned to face the music. A 727 crew made the same horrifying discovery midway between stations. They let down at reduced power and made it, but with insufficient fuel left to start the APU.

A ferry crew overlooked pitot heat, climbed into freezing clouds, was disorientated by incorrect flight instrument readings, and crashed. Fuel and pitot heat are not details; the subsequent pilot error verdicts cannot be argued, but it can also be argued that checklists were mitigating factors.

Our 727 checklist included 61 items requiring attention between gate and runway, all in the same size print. These ranged from fuel quantity and pitot heat to cockpit door (locked), rotating beacon (on) and landing lights (as required). Reciting responses exactly as published was emphasized. When the engineer said, "Clocks and altimeters?" the acceptable response was, "Check and set," and anything else annoyed inspectors.

But it was never stressed than only six of the 61 items were crucial: fuel, trims, flaps, speed brake, flight controls, and pitot heat. Get those right and you'd fly; the other 58 were foam on the beer. Another critical item was omitted: Were all the fires lit? We taxied out with the outboards to conserve fuel; the center stack of dead gauges could be overlooked when copying a revised clearance while expediting departure. It happened, fortunately in aircraft light enough to lift with two-thirds of normal takeoff thrust.

Pilots flying the North Atlantic were given several route choices, these "tracks" being revised every 24 hours. A crew, after studying winds and weather, made their selection and away they went. When they drew within range of ground radar in Scotland, they received a rude shock. They were far off course. Waypoint coordinates had been correctly entered and all three INS sets had performed normally. One number on the computer flight plan had been overlooked—the date. They had flown yesterday's track. Such booby traps in preflight data are not uncommon.

The FAA's predisposition to punish for even minor infractions is enough to warp a realistic sense of priorities. With good name and career at stake, a pilot's concern over staying legal can cloud his judgment. He is inundated by irrelevant technical data, inane reminders, ambiguous rules, and admonished to go fly by the book.

Rare shot of Braniff DC-3 in "new" paint scheme introduced in 1954.

In fact, the "book" is a small library of aircraft and policy manuals, company and federal regulations, charts, bulletins, and checklists, most of them revised weekly. The typical airline equipment manual is disorganized, badly written, and loaded with inconsequentials. A washing machine comes with better instructions. (The military handbooks I got were superb by comparison.) Company and federal regulations are vague and can be interpreted to suit the authors. The thrust is less to help pilots operate safely than to help ground people avoid blame when there's a trouble. The "book" can be summed up, "It's your baby, Captain."

A "book" writer's priorities differ from a pilot's. For example: By law every airline cockpit crewmember must have a "readily available flashlight in good working order" (FAR 121 excerpt), yet I've never seen that mentioned on a checklist. Once you've climbed into a night overcast and suffered a complete loss of cockpit lighting, you'll never again regard your flashlight as a detail.

If the half dozen "killer items" appeared in large red print as final pre-takeoff checks instead of being sandwiched between reminders to turn on transponders and fasten shoulder harnesses, there might be fewer attempted takeoffs with flaps up or trims out of limits.

The "book" is a guide, not a bible. For example, a thinking pilot distrusts warning devices. The 727 manual says that if speed brake, stabilizer trim and flaps are improperly set for departure, a warning horn will sound when thrust levers are advanced. He reads that as, "is *supposed* to sound." The same is true of fire bells, gear horns, speed clackers, and flashing lights. The blacks and whites of desk-bound technical writers must be translated to apply in a gray world ruled by Murphy's Law ("Anything that can go wrong—will.")

My crowd of World War II vets endured copilot apprenticeships lasting 16 years. We were well into our 40s before we got the coveted fourth stripe. For all of their frustration, those years gave us an invaluable insight into the real world of flying. Not only were we involved in many thought-provoking situations, we heard a hundred bizarre tales. Same of our seniors dated back to airmail flying. Their logs included virtually every American military and civil aircraft flown between World Wars I and II. What they remembered, along with our own experiences, molded our philosophy. Hangar flying brought training theories into perspective.

Today's pilot advances too rapidly to acquire such lore. He accepts the "book" at face value and learns its shortcomings as he flies. Study and experience will in time reveal what is important and what is detail.

It is all too easy to become lost in details and lose sight of the overall picture. Pilots intent on completing every nitpicking checklist item have taxied into ground equipment, other aircraft, and onto active runways.

Absorption with trivial cockpit chores has led others to mistune radios, overshoot altitudes and clearance limits, and miss traffic that filled the windshield.

There is a time to fiddle with fine points and a time to concentrate on stick and rudder flying. Once the details of every situation are recognized, they can be put on the back burner and maximum attention focused on things that really matter.

It was my good luck to fly with many military and airline skippers, no two of whom approached the job in exactly the same way. They ranged from relaxed to the point of boredom to tense to the point of ulcers. Nearly every one was a competent airman; the remaining few could drive a saint to drink—if they didn't scare him to death first.

Any pilot with 25 spotless years on schedules obviously had something going for him, no matter how he came across. Despite their diverse natures, the pros shared a fascination with flying. They worked at it and relished its challenges. "It's like bridge," one said. "You play with the same 52 cards but never get the same hand twice." They also shared a keen sense of priorities not based on slavish adherence to the "book." They flew by their own book.

There was a 12-step procedure for smoke in the Convair 340, the last being, "Land as soon as practicable." Now and then during transition my instructor would say, "I smell smoke," and I would rattle off the 12 steps.

While doing steep turns we smelled smoke—the real acrid stuff—and automatically I began the ritual. "Forget that (expletive)!" he yelled, "Let's get this (expletive) back on the ground!" which we did with all due haste. Having saved our skins we saved our tickets by making sure of the details. An inspector might visit the cockpit to see if we'd done it all by the book.

50

Dear Alex:

I'M LOOKING AT A PICTURE of Captain Alexander Galunenko in the left seat of the world's largest airplane. I'd like to ride with him in that monster. We'd probably need an interpreter, though he might speak English. Americans travelling abroad are frequently surprised, and embarrassed, by the number of people who speak fluent English. Translator or not, I believe we would soon be on an Alex-Len basis. Pilots understand pilots.

As a kid I was entranced by all airplanes, most of all by the fastest and biggest. Back then "fast" was the Army's sleek P-35 for which its designer, World War I ace Alexander de Seversky, claimed a top speed of 300. Five miles a minute, think of it! "Big" was Boeing's Model 314 four-engine flying boat. It weighed 82,500 pounds and carried 40 passengers 3,100 miles nonstop at 165 miles an hour. Incredible!

I pored over their pictures in *Popular Aviation* and memorized the details. The idea of someday flying airplanes like them was pure whimsy. The years since have seen hundreds of admirable designs that were neither fast nor big, some of them superior to types that were, yet I never outgrew a fascination with speed and size for their own sakes.

Fanciful dreams sometimes come true. Mine certainly did. Before World War II was over I had flown the C-46 Commando, the largest twin-engine airplane flying, the B-17, B-24, and C-54, the jumbo of its day. But I was only half-satisfied. Taxiing between lines of exotic Lightnings and Mustangs made my mouth water. I sat in every Allied fighter from the Wildcat to the Spitfire, moving the controls, soaking up the smells, and never got a chance to fire up one. That was intensely frustrating.

After the war there was little left but weekend time in a Reserve AT-6—until a National Guard P-51 unit was established at our field. I applied. We want pilots, not truck drivers, they said. Fighter airmen are a haughty breed. I appealed to the CO, an ex-B-24 pilot who had flown on the Ploesti mission. He relented. If I would accept an assignment no one else wanted (mess officer), I could join and fly when aircraft were not needed for training exercises. Had he said latrine officer, it would have made no difference.

Talk about pure flying fun! It was like riding a slingshot. Eventually the pros tolerated my presence in their formations and with mixed success tried to imbue the basics of their spine-tingling trade. Flying a Mustang to its limits was exhilarating, gut-wrenching hard work that left me sweat-soaked and aching with exhaustion. It was a thrilling education.

The old hands ran me ragged. After 200 hours, I had two probables but had been shot down a dozen times in the process. Those boys had combat time in their logs. The pilot who can promote an affair with a seductive fighter prior to marrying an elegant if less glamorous type for career flying has well sewn his wild oats. What he learns from the coquette comes in useful with the wife.

So, at a tender age my boyhood ambitions were realized. When you can fly airplanes like those, you don't have to grow up. Never again would I fly the fastest, but airline work served up a nice selection of the largest. The C-54 (modified, it became the airline DC-4), a monstrous machine in its prime, was followed by the larger DC-6 and still larger DC-7C, surely the ultimate piston-engine airliner. You were somebody if you crewed that great ship.

Turbine power gave new meanings to "fast" and "big" in aircraft parlance. Today's fighter can climb straight up after takeoff at a velocity a P-51 couldn't match in a vertical dive. The revolution in transports has been as dramatic. The Boeing 747 is 12 times the weight of the trusty old DC-4 and carries 11 times the passengers two-and-a-half times as fast. Even having flown both of them, such numbers still impress me.

Boeing's announcement in 1966 that it would build a "very large commercial transport" was greeted with mixed emotions. It's too big too soon, said the same experts who a decade earlier predicted the 707 would flop. Indeed, with a 10-abreast seating for up to 490 passengers, the airplane's size and cost gave pause to even the most optimistic.

Recently a Boeing official reflected, "We bet the company on it." He meant literally. And, in varying degrees, so did the 25 airlines that ordered 126 at $23 million a copy right off the drawing board. They dared not risk being left behind if the jumbo succeeded. More than 1,100 have been delivered or are on order.

I went to Seattle to see it. Tom Cole, one of Boeing's enthusiastic public relations people, led the way into the wooden cabin mockup. You didn't go on board in the usual sense; it was more like entering a ship. Clearly, the airplane would be immense. At Everett, in the world's largest building, the first 747 was nearing completion, a sight I will never forget.

You'll never hear a bad word from me about good old Braniff Airways but I did regret that we weren't large enough to order the new marvel. It was a real surprise to later learn that we had ordered one and would operate it between Dallas and Honolulu. In fact, our N601BN would be the 100th off the assembly line.

There was no reason to think we'd order more so my only shot at it would be as copilot. That meant giving up my 707 captaincy and a reduction in pay but the temptation was irresistible. We trained with United Airlines in Denver and Los Angeles. Sitting up there 35 feet high took a bit of getting used to, otherwise the 747 proved a pilot's dream.

The big day arrived. Our bright orange 747 appeared on the horizon, made a low pass across Love Field and landed. After the usual shakedown and publicity hops, it was put to work on the Hawaii schedule. The round-trip required 15 hours of flight per day. Industry experts said we could never maintain that pace, but we did. Routine maintenance and inspection grounded the ship for a day once every six weeks, otherwise it rarely missed a trip. Eventually we did buy more and I moved to the left seat of the 747-100, -200 and -SP, so wild dreams can come true.

Lockheed's giant C-5A Galaxy first flew in June, 1968, the 747 seven months later. Their eventual maximum takeoff weights were 769,000 pounds and 870,000 pounds, respectively, making them the two largest airplanes flying, a distinction lasting 13 years. Then, in December, 1982, Russia flew its Antonov An-124, a C-5A look-alike, weighing nearly 893,000 pounds loaded, 11 tons heavier than the 747-400. It was publicly displayed at the 1985 Paris Air Show. Twenty are believed to be in service today.

Russia has a long-lived love affair with big airplanes, having claimed credit for flying "the world's largest" 57 years ago with its "Maxim Gorky," a six-engine, 60-passenger design. Unfortunately, an aerobatic pilot attempted a loop close by and 52 people died.

Word that Antonov was building the first aircraft to weigh more than a million pounds came as little surprise to history-minded observers. The An-225 Mriya (Dream) first flew in December, 1988. Three weeks later it set 106 records during a single flight. These included: takeoff weight 1,120,370 pounds; payload—344,576 pounds; closed course average speed—505 mph; maximum altitude—40,485 feet.

The An-225 incorporates a stretched An-124 fuselage, new wing, six engines, twin rudders, and a 32-wheel landing gear. It will transport the 154,000-pound Buran space shuttle on its back and deliver heavy equipment to remote areas of the USSR. Its size and capacity are awesome. At maximum takeoff weight—1,322,750 pounds—the An-225 is 101 tons heavier than the largest steam locomotive and tender ever built. Put another way, it is as heavy as 373 Cadillacs. It could carry 155 of them 2,800 miles at 480 mph.

I want to see the An-225. I wrote the Russian embassy in Washington and was given an address in Moscow. There was no response to that letter, so I've written Capt. Galunenko, explaining my interest and asking if he can arrange a ride. It's a long shot but many strange things have happened in Russia since *perestroika* produced the policy of openness called *glasnost*. There's no harm in asking. All he can say is *nyet*.

51

The P-51 Mustang

AT 40 INCHES OF MANIFOLD PRESSURE the Merlin roared, at 50 it bellowed, and at 61 it thundered. The ensuing trip along the concrete was like riding a slingshot. Almost imperceptible nudges of rudder kept us tracking straight, then the gear folded, the doors closed, and flight began with silken smoothness.

The nose rose smartly and we rolled until the wing tip pointed toward the runway we left seconds before. Uncomfortably strapped in a noise-wracked cockpit exuding the aroma of hot oil, 100-octane exhaust, and sweat, I looked down at the rapidly receding hangars and the years rolled away. Once upon a time I flew Mustangs whenever the spirit moved, which often it did.

We climbed at 170 old fashioned miles an hour on the antique gauge and leveled at 8,500 feet, well clear of an airway. After a steep turn through 360-degrees to clear us, the lean nose dipped and speed rapidly increased. Then it rose through the horizon toward the blue sky and the view above was of farms and green fields. We dived straight down and swept back into level flight to complete the loop. Lovely, absolutely lovely!

The next best thing to owning a Mustang is having a friend who does, and second best to pouring on the coal myself was to sit in back while he did. My friend is Jeff Michael, a senior captain with USAir and an accomplished aerobatic pilot. His AT-6 air show routine will hold your attention.

His latest pride and joy is a North American Aviation P-51D-25NA, military serial number 44-74976, painted to resemble a yellow-nosed mount of the 361st Fighter Group, 8th Air Force. The "-NA" identified -51s built in Inglewood, California; the "-NR" series built in Texas. The serial denotes the procurement year as 1944; in longer serials, the first number was dropped, hence the "474976" tail number. The FAA registration is N98582.

Strictly speaking, all this should read T-6 and F-51 but those were postwar designations adopted after the USAF came into being in 1947; my nomencla-

P-51 roars past Reno pylon at full power.

ture is apt for the day when these fine airplanes earned fame with the United States Army Air Forces and speed was measured in miles per hour.

A sharp-eyed observer could mistake it for a P-51H because of its elongated vertical fin and rudder, a postwar modification Michael intends to correct. Purist that he is, he dons an A-2 leather jacket, helmet, goggles, and oxygen mask for serious picture-taking. "This mask must have rubbed your face raw after three or four hours," he said. Yes, it did.

The black "armor plate" behind the pilot's seat is in fact plywood. The rear seat installed in place of the standard 82-gallon fuselage tank is necessarily low to provide head room. It is usually occupied by Jeff's wife, Anna, a USAir flight attendant. "When I head for the field she beats me out the door," he said. There are no controls or instruments in back but I could read the most interesting numbers over his shoulder.

His prize draws admirers wherever it goes, most of them aware of its wartime record but probably not of its strange conception. In early 1940, the war picture in England was grim. British purchasing agents came to North American with an urgent request: build us Curtiss P-40s under license, lots of them, and in a hurry. The young California manufacturer was headed by James H. "Dutch" Kindelberger who had learned to fly in the Air Service in World War I and worked for Glenn Martin as a draftsman.

When Martin's chief engineer, Donald Douglas, left to form his own company, Kindelberger followed him to the West Coast. There he worked on the

DC-2 and -3 before leaving to organize North American Aviation, Inc., in 1934. His initial creation, a low-wing trainer, was designed on Saturday night with work beginning on Monday morning. The Army designated it the BT-9 and ordered 82. The idea evolved into the AT-6 (SNJ/Harvard), the most widely used Allied advanced trainer of World War II.

"You can't pull a rabbit out of a hat unless you put a rabbit into the hat beforehand," was a Kindelberger axiom. The Mustang went into the hat in 1938 when he went to Europe to sell his trainers. His tours of British and German aircraft plants gave him ideas about fighter design. He countered the P-40 proposal with one of his own: we'll build a fighter to your specifications, but of our design, and complete the first airframe in 120 days.

Accomplishing in four months what normally took two to three years seemed preposterous, yet North American's rapid delivery of trainers led the British to place faith in its ability to fulfill any promises made. Yes, they would consider his suggestions. On a Saturday afternoon in April, 1940, Chief Engineer Edgar Schmued's design group went to work. Preliminary drawings were on Kindelberger's desk the next morning at 10 a.m. and airmailed that afternoon to the British Purchasing Commission in New York. The proposal was accepted and the real work began.

The engineering team labored around the clock in drab quarters above the president's office. "The Mustang materialized out of that smoke-filled room where we knew no hours," said one engineer. "The lights never went out and for days all you could hear was the rustle of paper, the sharpening of pencils, and the noise of men knocking out their pipes on waste baskets." Another remembered a casting rushed to Kindelberger so hot it burned a hole in his desk.

Seeking diving velocities approaching the speed of sound, Schmued chose a modified laminar flow airfoil, a gamble in that its exacting tolerances had never been adapted to mass production. From the start Kindelberger insisted on a product that could be mass produced and quickly dismantled for maintenance and shipment.

Sixty thousand man hours, 2,800 drawings, and 117 days after work began, NA-73X was rolled out on a landing gear borrowed from a trainer. It was a phenomenal achievement. An Allison V-1710-38 water-cooled V-12 engine rated at 1,100 horsepower was installed. On October 26, test pilot Vance Breese made the first flight. The prototype exceeded all expectations. On its fifth flight another pilot ran the tank dry and the engine quit. The subsequent wheels-down forced landing in a field resulted in NA-73X being damaged beyond repair but Kindelberger knew he had a winner.

The RAF named it the Mustang and used it for ground support during the costly Commando raid on Dieppe, France, in August, 1942. In October, Mustangs were the first single-engine fighters to strike a German target from England. Two were acquired for trials at Wright Field but the Army evinced little interest. For all of its spectacular performance, the Mustang was considered a "foreign design" built to another air force's specifications. The official policy, which hinged on the P-47 Thunderbolt and P-38 Lightning, was later called "a mistake" by USAAF chief General "Hap" Arnold.

The new fighter had one deficiency. Its Allison engine with a single-stage supercharger did not provide the high altitude performance required of interceptors. World War I ace Eddie Rickenbacker was among the overseas observers who wrote to Arnold. "Crossbreed the Mustang with the Rolls-Royce Merlin engine and you'll have the greatest fighter in the business."

With a Merlin V-1650-7 installed, the Mustang was a fighter without equal. Most importantly, it could accompany American daylight bombers to German targets and reduce a loss rate that was becoming intolerable. Merlins were built in Detroit by the Packard Motor Car Company.

The P-51's subsequent success prompted William Green to write, "The Mustang created records from the day of its inspired inception and ended the war as probably the best all-round fighter to be employed by any of the combatants," high praise indeed from a noted historian whose country produced the Spitfire. The RAF operated more than 700 Mustangs; the USAAF more than 14,000 P-51s. The most widely produced model was the P-51D, of which 6,502 copies were built.

After the war the reserves offered the only chance to fly. On weekends you paraded up and down the ramp and endured lectures and the scratchy old VD films everyone knew by heart. The only reward was an hour in an AT-6. One day a silver P-51D eased up just off my left wing and what a sight, that lean, gleaming, sensuous beauty tucked right in there. On its fuselage was "KY-NG." Its grinning pilot crammed on power and with a mocking roll left me rocking in his prop wash. One-upsmanship.

The 123rd Fighter Group, Kentucky Air National Guard, was uninterested. My Form 5 listed transports hauling freight—C-47s, C-46s, C-54s—but no fighters. No way, they said. The CO, a decorated B-24 pilot who had flown on the Ploesti mission, eventually relented. If I would accept assignment as mess officer, I could fly when a spare -51 was available. Because the hangar had no mess, no problem. (I had forgotten the two-week summer camp at which event I was responsible for feeding 475 ravenous men three times a day.)

So began a brief but memorable affair with a beguiling temptress no suitor forgot. First, "Go shoot some no-flap landings in the AT-6," they said, get used to a fast approach, and arrival. Then a cockpit check and they said, "Go fly it." The big prop turned lazily, the Merlin shook convulsively and settled into its characteristic crackling idle. It was put up or shut up time.

You pulled 40 inches for the mag check, no more or the tail would rise. I was more passenger than pilot on the first insane takeoff dash and a mile from the field before I got the gear up and power reduced. Recollections of that hop remain vivid. The -51 had delightful control response but rough handling threw you against the straps or slammed you down hard on the chute. A mild dive sent the airspeed needle walking past 450, zooms sent the rate-of-climb to the peg, but the exciting Mustang was quite easy to fly and tremendous fun once you adjusted to its power and speed.

In due time I was tolerated on training missions and assigned to fly the tail-end slot in four-ship elements. The regulars tried with partial success to pound into my multiengine way of thinking the rudiments of fighter aviation. They could fly the -51 to the ragged edges of its limits. On one solo hop I spotted a -51 low and ahead. Easy meat. I should have known better.

I was sliding into position on his tail and almost within range when he vanished. Moments later the mirror was filled with a P-51 close behind. Nothing I tried shook him; it was like towing a glider. He was the pro, I the amateur. Never try to beat a man at his own game. I flew fighters but was never a fighter pilot. There's a difference. My lack of talent was somewhat offset by enthusiasm. While we were urged to log at least eight hours a month, I was told to curtail my flying in the interest of conserving the squadron's fuel allotment.

An hour in that ear-splitting, stifling cockpit was more fatiguing than three in a plodding transport. I roamed far and wide on cross-countries, selecting a new destination each time. The 1,100 miles between Louisville and Corpus Christi were easily covered nonstop on internal fuel alone if flown at 30,000 feet or more where the Merlin came into its own.

The last mass flight in this country—and perhaps anywhere—involved more than 700 military aircraft in President Truman's "Inauguration Air Review." The noise made by that two-hour low level stream up Pennsylvania Avenue must have been deafening. Old Harry must have sighed with relief at the sight of my -51, the Number Four man of the last element of the last squadron of the last group in that long parade.

Looking over Jeff's shoulder and down the slender nose of his prize brought back those fun times. He slow rolled and barrel rolled, then swept around the pattern with the Merlin barking in idle and set us smoothly on the numbers. Once I did all that myself.

There are some who see an interest in old warplanes as a sickness. They deplore the restoration, display, and operation of such "killing machines" and, of course, the P-51 was expressly designed to destroy enemy aircraft and ground targets. They do not see it as a superb machine, a rung in the ladder of aeronautical progress and an important reminder of events too often forgotten.

Our son was born in the last year of World War II. He's a USAir captain, flying enthusiast, and history buff. Jeff took him for a ride. Afterwards he said, "Now I know what you meant. That was an experience I'll never forget."

52

Getting hired, then and now

TODAY'S AIRLINE NEW-HIRE wouldn't believe the way things were done 40 years ago. After World War II, thousands of ex-service pilots competed for a few hundred airline berths. The air travel boom was years ahead. If you had an engineering degree and 2,500 hours, 500 as left seat/four-engine, they'd keep you on file.

The rest of us mailed applications by the gross and waited, repeating the ritual as discouraging years melted away. If you hadn't landed a job by age 28, you could forget the airlines. Finally, a special delivery letter arrived from Braniff Airways: Come to Dallas for an interview. I fired up my P-51 (actually, the Kentucky National Guard held the title) and shut down at Love Field 150 minutes later. It was no time to dawdle.

The employment manager was surprised: "I thought I wrote you yesterday."

"You did," I said, "I want this job."

The assistant chief pilot asked about hours, college credits, and so on, but not to see proof. You were who you said you were, a comment on the times.

"Report next Monday," he said. After four frustrating years and a five-minute chat, I was back inside the airport fence.

Braniff was smaller than the Big Four (American, United, Eastern, TWA) but, with 37 airplanes and 1,400 employees, was ranked seventh among the original 15 "grandfather lines." Headquarters were in a modest brick building and collection of drafty structures used by the USAAF 5th Ferrying Group during the war. Maintenance occupied a large wood hangar built for P-38 modification. Our class of eight raised the pilot roster to 210.

We made three DC-3 landings each to satisfy the legal requirement for copiloting, and went on the line that afternoon. We had no ground school, no uniforms, no brain bags filled with maps and manuals, and little idea of what was expected. You learned the routine by keeping eyes and ears open. My first captain was a low-key southerner who sensed my eagerness and gave me most of the flying, injecting a suggestion now and then. It was the most fun I'd had since leaving the Army.

The only reason for a copilot was that he was required by law. He didn't get checkrides or recurrent training. He simply had to be there; a trip could not leave without him. (Army copilots were "commissioned ballast.") His involvement was entirely up to his skippers. They worked him hard, recognizing the traditional one-man crew concept was outdated, even if the management and CAA didn't.

Probation lasted a year, after which a man with problems rated a hearing. It rarely came to that; the occasional misfit simply left. If you were truly interested in flying and showed reasonable progress, you needed not fret about being canned.

Of course, you hoped your captains turned in good reports. They had forms to complete covering skill, knowledge of the aircraft and routes, attitude, and so on. My year was nearly up when I drew an apparently humorless skipper with a hard-nosed reputation. He made it clear my future was on the line, that his report would decide my fate, and away we went into four days of the foulest possible weather.

Nearing home on the final leg, he handed me the form, saying he never reported anything he "wouldn't say straight to a man's face." It was a disaster. I couldn't keep the airplane right side up without prompting, my gauge work was hopeless, my attitude was poor, and my personal appearance disgraceful. After the last question, "Would you let your family fly with this pilot?" he had put, "Never!"

"That bad, eh?" I said, wondering if it meant industry-wide blackballing.

"Just kidding," he said 10 minutes later. "Try this." The second report was in all respects favorable, in fact, complimentary. No one had warned that gallows humor was his trademark.

Finding work with larger lines was not always as simple. Some gave written tests with questions like, "Which wine goes with fish, red or white?" and, "Where are Harley-Davidson motorcycles made?" though how that separated wheat from chaff was not clear. And your FAA medical was not enough; you had to take a "company physical." The paper shufflers were moving in.

There was no deadwood on Tom Braniff's payroll. He left hiring to the chief pilot, had implicit faith in his selections, and boarded "B-Liners" regardless of the weather. His informally recruited cockpit troops proved second to none. The broad range of equipment and routes they flew with distinction for half a century was unsurpassed by any other airline anywhere.

It's all quite different now. Landing an airline job has become an art in itself. The object of the game is the same—to join a seniority list the quickest way—but the rules have changed. No matter how well qualified he or she might be, the pilot who sends off applications and watches the mail is wasting precious time. It is almost impossible to comprehend the fluid job picture in today's huge industry without skilled help.

The best bet today is to contact an employment service like Future Aviation Professionals of America (FAPA). They know who's hiring, who's not, and what each airline wants. These agencies hold seminars at which hopefuls are taught how to write resumes, fill out applications and conduct themselves at interviews—that is, what to bring, the questions to expect, and how to answer them and even what to wear (for the men, two-piece suit, white shirt, conservative tie, etc.). It's nothing like when I walked into Captain Bill Walner's cubicle in a sweat-stained flying suit.

The fee for such coaching is money well spent. Incidentally, airlines respect the screening conducted by a reputable employment agency and listen to its recommendations. Hiring procedures today range from the sublime to the ridiculous. Some things never change. One interview might be enough. It's more likely that a candidate will confront a panel of pilots who know exactly what's required and another panel of desk-flying experts who think they do. All of them must be favorably impressed.

And, no matter if you've been flying President Bush around in Air Force One, you'll not escape writtens to see if you've got enough of the right stuff to copilot a DC-9. Expect a battery of psychological, math, vocabulary, and mechanical aptitude questions. You have to play their little game and remember—should they still ask—that white wine goes with lobster and Harleys are built in York, Pennsylvania.

A simulator check is often part of the "selection process." Prior experience is taken into account; a 172 pilot can get a nod as readily as a Learjet ace. Throughout the interviewing and tests, an applicant should remember there would have been no invitation in the first place had he or she not been a strong possibility. But it is hard to be optimistic and not feel relief when it's over.

Previous flying experience might count for naught during the next phase—training. When Braniff ceased operations in 1982, a number of our pilots found employment with other carriers. It meant starting all over again because seniority is not transferable.

A friend who had logged 6,500 hours with Braniff as a Boeing 727 flight engineer, copilot, and captain wrote that he was sent back through the entire engineer training program as though he had never heard of the Boeing. Five years later he requalified as captain, but only after completing the lengthy ground school/simulator/flight training required of first-time candidates.

Of course, airline flying today is a far more demanding profession than when we took it up 40 years ago. Great care is necessary in pilot selection and qualification.

The casual approach to employment we remember turned up new hires at least as promising as those selected after intensive scrutiny today. Our class of eight was typical of the pilots hired 40 years ago. One resigned and one died while deadheading aboard a trip that crashed. We remaining six flew 160,000 domestic and international hours in 14 types of aircraft from the DC-3 to the 747 without a single mishap or violation.

We were a tough act to follow if I say so myself.

53

Silver eagles

YEARS AGO IT WAS DECIDED to form an association of our older drivers. Membership was limited to retirees who had flown with us or predecessor lines and active pilots over 50. Widows became associate members. Most carriers in business 25 years or more have such retirement groups.

What would we call ourselves? Legend has it that "Bald Eagles" was a proposal withdrawn when a wag inquired, "Will the first word be four letters or six?" Anyway, "Silver Eagles" was chosen, by-laws adopted, and officers elected. The Braniff International Silver Eagles (BISE, spoken as "Byesee") were in business.

A quarterly newsletter was published by the lady who edited our company house organ for 28 years. The sole voting member who was not a pilot, she kept us up-to-date with letters from far flung members, snapshots of then and now and, as the years roll by, obits—too many of them. The highlight of the year is a convention each fall.

When our line failed you might think BISE collapsed. Instead, what began as a social club overnight became a service organization. While employees bore the brunt of the trauma, retirees were dramatically affected. We learned our medical and life insurance had lapsed; worse, there were rumors that pension funds had been milked into insolvency.

BISE committees tackled these and other questions and got the answers we needed. Volunteers worked tirelessly, answering calls from members all over the states and in other countries. Legal representation in bankruptcy court was hired, new medical coverage adopted, and factual information provided about the pensions. Their hard work restored sanity to a chaotic situation, particularly for those of us who lived far from Texas.

All of which was heartwarming, though not surprising. The bond developed between us while flying the line was no less strong now that we rarely saw one another. Only the problems had changed. I for one needed reminding that our brotherhood was alive and well.

The 18th annual convention was held in October, 1987. I had attended one years before while still active. It was time to go again, to meet old friends, and to thank those who had kept us alive during the rough times.

The festivities began on a Friday morning with a golf match. For the rest of us, hospitality hour was scheduled for takeoff at 4 p.m. My wife and I arrived at 4:05; the room was already crowded; our reputation for on-time performance was being maintained. (Once we offered a dollar a minute if we were late to each passenger; name another line to do that.)

It was like opening an old family album. Each familiar face was remindful of a ground school endured, or hours logged together for weeks or months, of incidents, some humorous, some sad or a bit hairy, of a chapter of airline flying that is almost quaint in retrospect.

I was buttonholed by a sprightly octogenarian who, aware of my interest in history, recalled his graduation as an Army pilot in 1927. He remembered the talk given his class by General Benjamin Foulois, one of the Army's original three pilots. He talked about flying our line's Lockheed Vegas, 10A Electras, and DC-2s, this old-timer who has witnessed airline history from airmail days to jumbos and SSTs, more than half of it as a participant.

Listening to him reminded me of the many early types flown by our seniors. They could tell you about the Jenny, DH-4, P-6E, F4B-4, F3F-1, B-10, and most of the other exotic military ships I read about as a schoolboy. A few began in World War I. One of our earliest chief pilots soloed in a French Caudron G-3 and logged time in a score of Allied and captured German types. A prize in my library is an autographed copy of his privately-published diaries. One BISE *Newsletter* reported his recent passing from the scene. His breed has all but vanished.

Virtually all American aircraft flown between the wars can be found in the logs of our older retirees. Few pilots today can describe anything built by such once-illustrious firms as Travel Air, Spartan, Timm, Alexander, or Waco. One of my skippers was Porterfield's test pilot; who remembers Porterfield? History buffs learn from books, but no picture or printed description compares with listening to pilots who flew those forgotten machines.

As a youngster, my special interest was airliners because I saw a remote chance of becoming involved in the infant industry. Airline employment put me in the cockpit with pioneers who had flown the Curtiss Falcon, Loaning Air Yacht, Fokker F-10, Stinson SM-1, Hamilton Metalplane, and many of the other types with which air travel was launched in this country. They would reminisce about their first days at any show of interest. A recent book lists 151 transports, domestic and foreign, used as American airliners. It's a fair guess that our pilots have—or had—flown three out of four of them, including the Concorde.

My (World War II) group added many types to the overall log. All of the better known U.S. fighters, bombers, transports and trainers were represented. One flew a tour on B-24s, a second on B-17s; his comparison of the two in combat revealed more than anything I've read. Such lesser known types as the F2A-3 Buffalo, F7F-1 Tigercat, B-32, and P-82 were included. The Hurricane, Spitfire, and Typhoon were flown by ex-RAF pilots who joined us.

The postwar era brought new hires with jet time, something only one of us could claim (he flew the P-59). We learned about everything from the P-80 to the B-58 and what it was like landing a jet on a carrier at night.

There are few places on earth that at least one of our pilots has not seen, the Polar regions included. Many of their experiences were remarkable, some almost beyond belief. Renewing old acquaintances brought it all back. We are not unique; BISE is like similar clubs on other lines. The names on our airplanes were superficial differences. We are members of a vast international brotherhood.

The banquet was on Saturday evening. About 100 pilots, their wives, and a few widows attended. After dinner the names of members who had passed away during the past year were read—George, Fritz, Bruce, and the others, leaving legacies best known to us. Recognitions and awards followed. The speaker began by observing that the pilots present had logged 325 years of flying. Had all our surviving members shown up, he could have said 1,500 years.

There could have been no finer career for a boy with the itch to fly than to work with such people. The art of flying attracted good individuals from the beginning. It still does. Its mystique has faded but the indefinable appeal remains. Pilots are restless; they are intrigued by the new and different. They seek challenges and accept risk as the price of achievement. They make things happen. A psychologist wrote, "They go out and meet life head-on." That is true of all the pilots I have known, no matter how they used their skills.

As a youngster I fell in love with airplanes. Flying promised excitement and escape from the dreary existence led by my elders (a naive conception embarrassing to admit even now). Its true nature was a surprise. The art required intensive study, hard work, was in some cases hazardous, and in most cases underpaid. As enviable a pursuit as airline piloting seemed, it was—and is—frustrating and stressful to a degree that laymen will never comprehend. Yet, I believe my fellow retirees would agree that the rewards far outweighed the drawbacks.

The convention talk was about where we live now and what we are doing, about kids and grandkids, about today rather than the past. Curiously, there were few recollections of airplanes or places or experiences. Flying, the powerful magnet that once drew us together, was scarcely mentioned. What we missed was the comradeship. What we missed was each other.

54

Keeping up

CALLED THE "AGING PROCESS," it is the inexorable progression of subtle mental and physical changes we all undergo as the years roll by. For professional airmen whose careers depend upon continued well-being, it is a matter of concern and, in too many cases, worry.

Someday it might be possible to more closely measure the effects of aging on an individual basis, but studies to date have produced no more than general conclusions. Our physical and mental powers increase and decline at different rates and therein lies the riddle. Some people are old at 40, others young at 60.

Unpredictable and undetectable deterioration compounds the problem of determining if a given pilot remains fit to fly. A month after his six-months physical, a 42-year-old friend in apparent good health slumped dead in his seat on final approach. A copilot in his mid-30s napped after lunch and never awakened. A captain nearing a midwestern airport suddenly was back in a B-17 over Germany and the other airplanes in the pattern were Messerschmitts. Such sad and mysterious events shook us.

Physical and mental problems ground professionals every year. It seems an occasional thing yet the cumulative effect is surprising. Looking back through a 10-year-old seniority list reveals far more dropouts than were remembered. Out of sight, out of mind.

The realist does not find this topic depressing for he regards his career as he does a flight. He takes off after thorough preparation and fully expects to complete his flight without incident. At the same time, he acknowledges that mechanical or weather problems could cause delays or even diversion to an alternate destination, none of which possibilities alarms him. They go with the job. He sees his career in the same light, not dreading trouble but prepared for it.

Airline flying provides an opportunity to observe pilots of all ages over a working lifetime and compare your situation with those of cockpit companions.

My first skippers were airline captains recruited for wartime transport flying in Africa; each one one was old enough to be my father, which upset the then popular notion that flying was a young man's game. Bald heads, bifocals and all, they flew like angels.

My first airline berth was as a DC-6 flight engineer. Because our left-seaters earned top money on night runs while copilots and engineers drew flat wages, the most junior copilots and engineers were teamed with the most senior captains. I could have been the youngest child of any of them.

Watching those patriarchs perform was a revelation. You could learn much about a pilot on a night run from Chicago to Dallas through Thunderstorm Alley in those pre-radar days. Those ancient mariners were more than a match for almost anything we encountered. Their prescience was uncanny. When old Clancy pointed to the mixtures and said, "Sweeten 'em up—and get the gear up off the locks," you could make book on a wild ride coming up. "Keep an eye on Sid's cigar," a copilot advised about another old-timer. "If it goes out and he doesn't relight it, cinch your belt."

Which is to say, the loss of youthful reflexes was offset by expert judgment gained through long experience. At least, such was the theory and it proved valid in most events. Now and then an old boy's sixth sense was all wrong and we paid dearly for it, but it was clear that most pilots could fly into their 50s and some into their 60s, by which time they were shrewd enough to avoid most problems and deal with the rest. (The feds would later decide that age 60 was enough, but that's another tale.)

The knack of quickly absorbing new ideas declines with age. We young engineers breezed through ground schooling. Senior copilots sweated. For skippers trying for the left seat, the exams were pure hell. Flight training followed the same pattern. The older the student, the more slowly he acquired proficiency in a new type.

Our first jet was the 707-227, a sizzling hot rod with control feels quite different from those of propeller types. Some veterans nearing retirement needed 50 hours of dual to get ratings. A few took sick leave upon completion of the course; they were truly exhausted. Others elected to finish up on familiar piston equipment at less pay rather than endure the miseries of more training.

Personally, each transition was tougher than the last, at least with respect to ground schooling. Once I got to the airplane I had it made. Flight training was hard work but interesting. In due time my classroom anguish would amuse young engineers. For most of us, the longer away from formal schooling, the more we had forgotten about studying. Concentrating on studies and tests is a skill in itself. One day it dawned that I had reached the age of the old fossils who had shown me the ropes 40 years earlier.

After flying with professionals for a number of years and considering my own feelings, it was obvious that an upbeat outlook is every bit as important as physical well-being. The individual most likely to enjoy a long career is one who likes the flying life. He takes his work seriously, but not himself. He does not imagine he is special but believes his work is and would not trade it for any other. He's sensible, adaptable, cheerful, and for all his grousing relishes the challenge of new ideas. He wants to prove to the young studs (and himself, of course) that he still can cut the mustard.

Professional piloting is not a job, but a way of life. The pro is a slave to the calendar and clock. His life revolves around the month's sequence of trips. To miss or delay a trip is unthinkable. His performance is more closely monitored than that of any other professional. The pressures are at times intense and on occasion almost overwhelming. Yet, despite its anxieties and frustrations, a flying career is exactly right for certain men and women.

Attitude and outlook change with age and circumstances. Personal objectives change. In the beginning flying is a totally absorbing adventure. A decade later it is seen in a quite different light. The 24-year-old fighter jock loves every minute of the life, but knows his is a young man's work. Bent on a lifelong flying career, he serves out his enlistment and is hired by a corporation or airline. He faces readjustment, to put it mildly. Nearly every time, he fits right into the new picture. After all, adjusting to new situations, some of which come as sudden surprises, is what flying is all about.

The development of a calm and patient attitude toward the flying life is not an effortless evolvement. It must be worked at. It is all too easy to dwell on the drawbacks and lose sight of the overall picture. No sadder individual is a pilot who has lost interest in flying to the extent he feels trapped by it. He can't afford to quit and doesn't want to go on. It happens. One must wonder if his lack of self-discipline would not have made him discontent in any work. Continuous thinking on the dark side can develop into a corrosive habit.

Certainly, the current instability of cockpit employment is reason for great concern. In some respects I think those of us who retired a decade or so ago never knew how good we had it. On the other hand, thinking back on the problems we remember, I wonder if today's professional appreciates how good he has it. The life today is not so much better or worse than it is different. Flying as a career never was and never will be what we might want it to be, but it surely wins out over whatever is in second place.

In the final analysis, any career is most often as rewarding as an individual makes it. The years roll by, adjustments are made, and the job remains as appealing as ever. A surgeon friend said, "A 17-year-old kid decided I would be a doctor, and I've been stuck with it ever since." In my case, a Boy Scout who had yet to sit in an airplane decided I would be a pilot.

I don't know about the teenager, but the scout was right. I had a blast.

55

You can go back

SEVEN YEARS AFTER SETTING BRAKES in Fat Al for the last time, I was back in that lofty cockpit, conscious of the muted rumble of idling JT9Ds, looking down a long runway, awaiting clearance to roll. It was unbelievable!

"Check list complete, Captain," said the engineer. Then a final glance at the killer items: trim, flaps, fuel, spoilers, flight controls, pitot heat. An old habit, it takes 10 seconds and is cheap insurance.

"Cleared for takeoff," said the tower. I eased the thrust levers off the stops, waited for the rpms to stabilize, then worked the levers forward until max power stood on the gauges. The huge Pratts growled and we moved, all 375 tons of airplane, fuel, passengers, and freight and accelerated—much too slowly to outward appearances. You sit way up there and seem to be doing 50 when airspeed confirms 100. Remembering that, I regarded the remaining runway without apprehension. A touch of rudder kept us entered.

The red lights ahead drew steadily closer. The most interesting two dials read 130 knots, 140, 150. The end of the runway was in clear view. "V-1," said the copilot, then, "Rotate." Ease back, haul the nose up, but not too high, and feel the slight thumps of four main struts extending. "Positive rate of climb," said the copilot. "Gear up," I said. At 200 feet a red light winked out; all 18 wheels were tucked up and their doors closed behind them.

We labored up at 500 feet a minute, decreased the angle of climb at 800 feet and, with speed inching higher, retracted flaps on schedule. A minute later we were clean and riding with rock solid smoothness. "Climb power," I said. "You've got it, boss," said the engineer, pressing a button behind the thrust levers.

Good show. The most critical aspect of flight was history. Now to work our slow way up to cruising level and set course for . . . London? Paris? Were we leaving Guam for Hong Kong? Or would we skirt the Russian coast between Anchorage and Tokyo?

A Northwest 747 flares to land at Honolulu.

I was getting carried away. In fact, the other crew seats were vacant and I was going nowhere. The runway was an image on a screen and the realistic sounds, motions, and readings on the broad panel were computer generated, the whole adventure being reenacted in a simulator. But the illusion was near perfect and the next best thing to actually flying a 747 again.

Behind me sat longtime friend, Jack Heggie. We go way back. When I was a lowly DC-6 flight engineer, he was a senior copilot. When I made copilot, he made captain. When I made captain he was an instructor and check airman. In these capacities we logged many miles together. For years the Heggies lived two doors away. Ours is one of those warm and enduring associations that are common in the flying fraternity.

After retirement, Jack remained in simulator work, instructing and giving checks in the 727 and 747 to crews of a number of carriers. Recently Jack and Sara came to visit. When retired pilots get together there is little talk of flying but, during a lull in the conversation, I carelessly remarked, "Give me a week to hit the books and five simulator rides and I could requalify."

Jack shook his head. "You could do it in two."

"No way," I said, "it's been too long."

"Why don't we find out?" he said, and the experiment was arranged.

So there we were, lifting from the pretended airport and climbing through make-believe cloud to a simulated flight level, all of it inside a four-story building. From his panel, Jack played copilot, flight engineer, and air traffic controller. He could dish up fair skies and smooth air—or engine failure, smoke in the cockpit, a flap malfunction, wind shear on short final, any emergency or combination of them. There is no end to the unpleasantness that can be programmed into simulated flight.

When I strapped in and looked around I felt right at home. The feel of the controls and thrust levers and the spectacular view over the nose were familiar; my last trip could have been yesterday. The broad instrument panel raised few questions but the array of overhead switches was a mystery. You face the forward gauges hour after hour and look up only when you need something, like the seat belt sign or to unlock the cockpit door. Some study would be needed to bring all that back, not to mention the procedures and limitations reviewed in every pre-checkride oral.

He made it easy; the clearance was to climb straight ahead to 5,000 feet; there was no abort at V-1 or engine loss at rotation. We reached the assigned altitude with the blue side up and after much fiddling with the power, during which I drifted off altitude, I nailed down 260 knots, the minimum maneuvering speed for our weight. My instrument scan was poor.

"Try a 360-degree turn to the left, 30-degree bank."

That went fairly well, as did steep turns each way, though airspeed once sagged 15 knots, an unacceptable deviation. The recoveries from approaches to clean and dirty stalls weren't too bad. The aim is to resume speed without altitude loss. I managed that but over-controlling produced secondary stick shaking and that's not the way to do it.

The hand flown approach to minimums on four engines was a piece of cake. I remembered the drill; ease power to idle at 100 feet, start a mild flare at 60 feet, then listen to the altitude call outs and John Wayne it to touch down. "Thirty feet, 20, 10, and we were down. The speed brake handle motored back while I pulled them into reverse and braked. We stopped with lots of room left on the Disneyland runway.

The scene ahead changed; immediately we were back in takeoff position. That's the beauty of flight simulation. There's no time wasted, no aircraft operating expense and most importantly, no damage from a badly botched arrival. In the very beginning we learned and practiced all normal and emergency procedures in the airplane itself, usually in the wee hours, and there were more training accidents than the industry cares to remember.

We hammered down the runway again. "V-1 . . . Rotate." As the nose lifted, Number Four expired and I was extremely busy maintaining heading and a positive rate of climb. The seven years began to tell; my wallowing course was anything but the precision flying expected of an ATP. But I retrimmed and kept it out of the trees while being vectored around the field for an approach, working up a fair sweat in the process.

"Enjoying yourself?" said Jack.

"The best part of this is knowing that my income doesn't depend on the outcome," I said.

Instrument scan is the ticket; mine had gone to pot. By the time I noticed heading and altitude, they were out of tolerance. We broke out at 100 feet to the left of the centerline and a bit high. We could have made it but a truck was approaching the runway at mid-field. Before simulators with visual displays, they'd say, "There's a truck on the runway, take it around!" Now you see the truck and pray it will hold short, which it never does. Sure enough, this idiot kept coming. Max power, stand on the left rudder, positive rate, gear up, retrim, and climb back into the overcast. "After takeoff checklist," I said, trying to remember what to do next.

Finally, a nonprecision approach to the downwind leg at 500 feet. Jumbos rarely make circling approaches but it's an interesting and demanding procedure. My execution was sloppy and we arrived with a shameful impact. In the real world I would have remained in the cockpit until the passengers had gone.

They say you can't go back. They mean you shouldn't because you will be disillusioned. But now and then you should, just to see. My good friend gave me a rare chance to learn what I remembered of a wonderful time in the past. I'll never fly a 747 again—but I could, and that's nice to know.

Thanks, Jack.

56

Plunges and air pockets

TELEVISION REPORTS on the Gulf War provided a revealing glimpse of American newsmen at work. Video cameras put us on the front row at briefings. This let us compare official statements with subsequent press and television news stories. Reporters don't like this arrangement. Testifying before a congressional committee in February, Walter Cronkite said, "Briefings should not be covered live." His objection is understandable considering the inane questions asked by some of his young colleagues such as, "General, when are you going to begin the ground war?"

Briefing officials were repeatedly badgered for sensitive information. Reporters displayed an appalling ignorance of warfare and the necessity of secrecies. It was an embarrassment to watch. Foreign correspondents asked more sensible questions and got informative answers.

There was wry humor in this for us airline people, knowing the public was seeing firsthand how news is gathered. And a tinge of envy since, unlike the services and government, we cannot impose sensible limits to details when we make the headlines.

The dramatic growth of civil aviation during the past half century came about despite inaccurate and lurid reports of accidents. The travelling public came to appreciate the minimal risks of flying and long ago abandoned trains and ships. Personal and corporate flying are not novel. Newsmen know this; their work requires air travel. Yet their sensational accounts when things go wrong can turn a layman's mild apprehension into fear, causing damage to our industry. Why do they do it?

Think money. The media's professed aims are to report events, record history, and present both sides of important issues, but the bottom line is none of the above. The media primarily exist to earn profits for their stockholders. Success is measured by circulation and ratings that in turn determine advertising rates. Audit Bureau of Circulations (ABC) and Nielsen numbers add up to black ink or red.

A thrill-a-minute approach is the result. There must be exciting stories to tell even when nothing exciting has happened. Unless headlines and newscasts rivet attention, ratings slip. Non-news is often dramatized beyond recognition and real news embellished beyond reason.

When an Eastern 1011 momentarily lost power, it "plunged" toward the sea. When a KLM 747 flamed out in volcanic ash, it "fell 13,000 feet." It took a "frightening plunge," by another account. A child passenger was bruised when a 767 encountered an "air pocket," after which "the pilot" made an "emergency landing." The front page of our paper carried a picture of a Cessna 150 on its back. Nose gear failure, no injuries. Inside was mention of a car collision that took three lives. No picture. Such is the media's sense of newsworthiness when it comes to aviation.

Lack of technical background doesn't faze a creative reporter. One described wind shear as, "A strong gust from behind or to the side of an airplane that can reduce its airspeed to a level at which its engines will fail and cause it to plummet to the ground."

Another came up with, "A sharp change in wind speed and direction which can reverse the flow of air across a plane's wings and cause a loss of lift." Now you know. We might smile at such gibberish but it frightens passengers.

Accidents depress insiders. Aware of the work, training, and meticulous attention to detail that make flying safe, we are deeply hurt when things go wrong and dread the headlines sure to follow. Certainly, a bad crash is big news. Everyone wants to know who, where, when, and why. The first three are easy; the last might not be known for months, a fact reporters will not accept no matter how often the complexity of accident investigation is explained.

They prowl the scene, hound investigators for the one answer none can give and otherwise get in the way. Frustrated, they resort to speculation based on rumors and hearsay, indifferent to the unfair damage caused. A classic example of irresponsible reporting followed the crash of USAir Flight 5050.

On a rainy September night a USAir 737-400 began rolling at LaGuardia Airport with the copilot at the controls. The thrust levers automatically advanced. The copilot inadvertently disengaged the autothrottle system, whereupon the captain manually moved the levers. The airplane drifted to the left. Attempts to correct with rudder and nosewheel steering were unsuccessful. The captain took control and aborted. The 737 slid off the runway and crashed into the approach lights. Two passengers died.

The captain, copilot, and a deadheading Pan Am pilot assisted in the evacuation of survivors then left to consult with Air Line Pilots Association lawyers, the customary legal procedure. An airline pilot need not answer questions until he has met with union representatives. Unable to locate the crew, reporters hinted there was shameful conduct to hide. "Hit & Run! Pilot Goes AWOL," headlined the *New York Post*. "Pilots Duck Crash Probe," declared *New York Newsday*. Even the staid *New York Times* ("All the News That's Fit To Print") said the crew was "unavailable without explanation."

It was rumored the captain had "mumbled irrationally" prior to takeoff, untrue but published. A National Transportation Safety Board official mentioned the inadvertent autothrottle disengagement. Reporters dashed off "wrong button" stories. This minor event was mushroomed into a *Newsday* headline, "Feds: Pilots Really Blew It."

It was reported that the captain was on one of his first 737 captain trips (true), but not that he had logged more than 2,500 copilot hours in 737s, half in the –400. It was said it was the copilot's first 737 trip (true), but not that he had considerable captain experience with a commuter line and had undergone extensive training at USAir.

Then, "Crash Pilots May Have Been Drinking" headlined the *New York Post*. An unidentified source told a policeman he had bar-hopped with the crew the night before. Company officials denied this, offering proof, but the story ran anyway. The truth was both pilots were between trips and in the LaGuardia terminal throughout. No retraction was printed.

The impression left by this "news" was that USAir dispatched a trip in the hands of an inept crew. The captain was inexperienced and perhaps mentally unstable. His "novice" copilot was an all-thumbs rookie who "pushed the wrong button," causing the crash. Both pilots might have been drinking, which could explain why they fled the scene; they were fearful of tests for alcohol and drugs (which were not required). What other conclusions could be drawn?

No commercial enterprises are more closely regulated than airlines. Every aspect from aircraft design and construction to maintenance, operation, and even bookkeeping follows strict rules and is closely monitored by the federal government. Everyone involved is accountable.

The media are self-regulating and accountable to none but themselves. They are free to observe and interpret events as they choose. Too often they mock the First Amendment. The approach is slapdash with little regard for accuracy and fairness. To seize attention is the thrust rather than reporting facts responsibly. News is perishable and deadline pressures are intense, but this is no excuse for what we have endured in times of crisis.

Television is the most important means of communication since the printing press. When it comes to a crash, the printed word can't match it for excitement. We see the smoking wreckage while listening to the last words of pilots trying to avoid disaster. But is this news or ghoulish entertainment? Never mind, it's what the public wants.

Or is it? Television has proved a double-edged sword. It shows both the event and its reporters. A CNN poll taken during the Gulf War found viewers were not altogether pleased. Seventy-eight percent thought the military was telling all it could, 79 percent favored press restrictions and 57 percent thought the military should be given more control of reporters. Our low regard for the media is more widely shared than we knew.

Reader/viewer dissatisfaction goes beyond TV, according to a *Los Angeles*

Times Mirror five-year study. It found public confidence in the news media "as an impartial observer of the political and social scene" is eroding. Responsible newsmen also are bothered. "Many members of the press community also confess considerable unease and significant self-doubt about press performance," says the study.

This is a hopeful sign. It's time for creditable journalists to restrain the headstrong incompetents in their ranks. They can do better than plunges and air pockets.

57

Career flying in the 1990s

HOW INTERESTED IN FLYING are today's young people? Has a flying career the same appeal as when I was a teenager? For that matter, is it as desirable as I found it to be?

I must have flown 10 years before logging as many hours in airplanes as I had spent watching them. Only bad weather canceled my after-school pilgrimages to Bowman Field, Kentucky's prime facility. Bowman was a square mile of grass with a small brick terminal and three hangars. The dozen civil and Army reserve types based there rarely came to life except on weekends, so it was not unusual to wait an hour for any action. In the meantime the field was as quiet as a graveyard.

Then I'd hear a faint drone and a dot low in the sky would become a Waco or Stinson or a huge Eastern DC-2 and simply watching it land amply rewarded my patience. It was not mere curiosity that drew me; flying was an obsession.

Those lean years in the 1930s are often called the "Golden Age." It certainly took gold to get inside the fence; private flying was sport for the wealthy. Airline travel was expensive and perceived as adventurous but risky. Military aviators were reputedly the cream of the pilot crop. Aviation had yet to earn a true identity.

The war changed everything. Clearly, the superman air about the military was undeserved. High school graduates by the tens of thousands were quickly trained to fly aircraft far more demanding than anything in the prewar services. Apparently, many people of average talents could become pilots. The reliability and safety of wartime transport operations hinted at an exciting future for commercial air travel.

But first, old perceptions had to be changed. A book could be written on the success of airline advertising in the decade following World War II. New images were created despite sensational coverage of crashes in the press (which all the while accepted a major share of airline promotion dollars). Air

Western 707 skipper begins his day's work.

travel was portrayed as dependable, comfortable, and cost competitive considering the time saved. By 1950 the "air age" had dawned.

The private airplane was promoted as a useful adjunct to the automobile. Some builders went as far as to promise "a plane in every garage." Pleasing colors and car-like appointments suggested that the transition from highway to airway was no problem.

It was rarely said that flying for its own sake was a stirring experience and downright fun to boot. That would never do. The thrust was to deglamorize it, to present air travel as a routine public service, and private flying as the chic way to make business trips. Was flying desexed in the process?

"Not at all," said an official of Future Aviation Professionals of America (FAPA). "The airplane is still fascinating." While FAPA's main function is bringing professional pilots and employers together, it receives inquiries from young people wanting to learn about flying careers. Some 450 of its "Career Pilot Starter Kits" outlining basic requirements, training costs, and employment prospects are sold a month.

"That indicates considerable interest," he said. "The youngster who enjoys the skill of driving and likes machines is naturally interested in what the airplane can do. The historic allure might be gone, but what you can do in an airplane hasn't changed much. The fascination is still there."

Most young clients want to know about the airlines, he says, because they are the most visible, but some look to the services. "They've watched the Blue Angels and Thunderbirds and seen 'Top Gun' and so on." Others are interested in corporate and other aspects of general aviation. He notes that while pilot starts are down, the trend is that more students are entering for the express purpose of becoming professionals and sticking with it until they qualify.

Of course, today's youngsters are not as smitten as my generation or, at least, not in the same way. Flying for us was largely a mystery, its future uncertain. The mystique that overwhelmed us evaporated long ago. There's no more gee whiz aura to being a pilot. Yet, as commonplace as it has become, the operation of a fast and complex vehicle through space without regard to terrain is sure to attract the attention of anyone with the least imagination. The challenges involved in modern flight are obvious.

There is more to the life than technical skill. The variety of personal associations, the travel, irregular hours, and continuous influx of new equipment and procedures produce a unique environment. For the individual with the temperament for it, there is nothing else as satisfying. "People who get into professional flying rarely leave, and those who do, come back," said the official. Flying is not simply a job, it's a way of life.

"It doesn't have the stature it did," he said, "but it remains a profession with a great deal of respect. Some people will make more money and others will have more time off, but few will earn the same money and have as much time off."

How has the life been affected by the upheaval of the 1980s? "It's not like it used to be," says a 22-year airline captain who has survived a bankruptcy and a merger. "It's much more stressful. I don't mean the traffic or weather or checkrides—I can handle all that. I worry about the future. Will I keep what I have or is some hotshot corporate raider going to blow it all away? All of us are worried and we're weary of thinking about it."

What about the new hires entering the picture? "I don't see the same enthusiasm I had in the beginning," he says. "They are sharp, a good group, and the quality of their work is excellent, but all they talk about is who is merging and how they might be affected. The world has been turned upside down. Nothing would surprise me." Although employed by a major line that seems sure to survive the shake out, he says, "The chances are I'll be wearing still another uniform when I retire."

While airline pilots are the most visibly affected by deregulation, the frenzy of collapses, takeovers, and mergers throughout business has been felt by corporate and other general aviation people. Flight departments have been expanded—or abolished. Some pilots advanced, others suffered drastic reductions in pay and benefits, or lost everything and had to begin all over again.

"It all depended on where you were sitting when it hit the fan," said a pilot with a large carrier that has experienced rapid post-deregulation expansion. Less fortunate colleagues were disillusioned and embittered. One 727 captain who found himself unemployed in mid-life told a student pilot to "forget flying, get into something else."

Many insiders believe the tension will ease and the profession will slowly recover the security and stability it once enjoyed. Their optimism appears well founded. Barring a major economic slide, the volume of air travel will steadily increase in the 1990s. The megacarriers now emerging will have dramatically increased fleets. Hundreds of new airliners will be delivered during the next decade, each requiring five to seven crews. Some of the retired equipment will be converted to charter work or freight hauling here and overseas.

All of which points to a hiring surge at the very time pilot retirements (mandatory at age 60 on scheduled carriers) are nearing 2,000 a year. The law of supply and demand works both ways. Wage and benefit disparities enforced during the pilot surplus of the early 1980s will be quickly dissolved if a serious shortage develops.

Meanwhile, many pilots watch, wait and hope for the best. "It's still a good job—if you work for the right company," said the captain who worries about the future.

That reminded me of a question I dared ask a top skipper when I was a new hire, "Is this a good airline to work for?" After a pause, he said, "That is something you won't know until the day you retire." He was close. It was a good company. I flew until normal retirement at age 60. Seven weeks later the airline folded.

58

Care for the aged

IF BY SOME LOVELY QUIRK I came into some extra money, I'd restore old airplanes. I mean a nice sum, say a million after Uncle's clip. Restoring old airplanes is expensive; you could quickly run through a million but, with care, much could be accomplished.

For some there is real satisfaction in saving from extinction things worth preserving—an old house, car, locomotive, clock, furniture, books, toys—the list of things that people preserve is endless. Fortunately, some collectors are intrigued by old aircraft. Unfortunately, preserving for history's sake is not always the motive.

At an air show I watched an immaculate old biplane being flung through aerobatics its designers never intended when it was built in the 1930s. It was a display of the owner's nerve rather than his antique—but it *was* his airplane. It survives in a museum; he was less fortunate.

Since World War II an appalling number of meticulously restored aircraft have crashed, not to mention the loss of life in many cases. The day we arrived in England two vintage jet fighters collided at an air show. The evening BBC news showed them going straight in, one with two men aboard. A week later, on the day after we left Plymouth, a PBY reenacting the NC-4's arrival there 60 years earlier was wrecked on landing.

You'd never see a restoration from my dream shop doing rolls at 500 feet. Fly it? Yes, but very carefully, and I'd pick the pilot. Of course, if it were the sole remaining copy, we might fly it once for pictures. In my book, owning a piece of history confers certain responsibilities, no matter whose name is on the title.

Until recently the world aviation community showed little interest in its heritage. The emphasis was on tomorrow; many of yesterday's notable flying machines have vanished. No RAF Sterling, Hampton, or Whitley survives, and we have our own losses to deplore. And much priceless memorabilia has been lost. A directive concerning the fire hazard of stored celluloid led one military base to destroy thousands of negatives of extinct Air Corps types. Numerous sad tales are told by air-minded collectors.

Traditional historians smile at the idea of "aviation archaeology" and cannot fathom the National Air and Space Museum attracting more visitors than any other museum anywhere. An engine exhumed from the crater it dug 50 years ago can be as interesting to the air history buff as 500 B.C. pottery is to the student of ancient history. Members of The International Group For Historic Aircraft Recovery (TIGHAR) are searching for an airplane engine that won't bring $25 as scrap metal if found. But its discovery would solve a 60-year-old mystery.

On May 8, 1927, a large biplane left Paris, dropped its wheels, and headed for New York. The pilot was famed French ace Charles Nungesser, the navigator, Francois Coli. Never seen again, they "vanished like midnight ghosts," in the words of Charles Lindbergh who 10 days later left New York eastbound. There are reasons to believe the Frenchmen made it to this side, ran out of fuel, and crashed in northern Maine.

So far, 10 "Project Midnight Ghost" expeditions, financed in part by corporations and government agencies, have been unsuccessful. The search continues. "We don't expect to find much more than the engine," said TIGHAR's Ric Gillespie. What then? "It, and any human remains found, will be flown in a French airplane to New York. Symbolically, Nungesser and Coli will have completed their flight. Then they will be flown to France, the engine for museum display and the remains returned to the families for burial." TIGHAR's members, he says, "are sparked by a compulsion to know the truth, to solve the mystery."

The 250 TIGHAR members hope to recover the only B-17E known to have survived the war. Damaged in combat, it bellied into a Papua New Guinea swamp in 1942 and is virtually intact. Despite diplomatic problems and the considerable expense of removal, Gillespie is hopeful it will be returned, restored, and displayed in an American museum. "We will not own it," he says. "We are a non-profit foundation and cannot own anything. Our sole interest is in seeing historic aviation artifacts responsibly preserved."

Profit has been the motive behind many restorations. What more interesting ways to earn money can there be? Being a day late and/or a dollar short cost me more than one chance to get involved. In 1950, a faded P-51B sat near the terminal at Wichita Falls, Texas. During our 20-minute stops I'd admire it. The owner said he'd put it in shape and fly it to Dallas for $3,600—but that was a year's copilot wages. Before taxes and two kids. There was a BT-13 for $450, an AT-6 for $800, a C-78 for $1,200, and not a prayer of buying any of them.

Remembering the Curtiss Jenny almost makes me cry. Bought surplus after World War I, it was stored in a warehouse in our small rural town. We schoolboys gazed at it in wonder; it was the only airplane any of us had ever seen. Forty years later I called a friend whose father had owned the warehouse. "You're six months too late," he said. "It burned to the ground. Had I known, I'd have given you the airplane."

Keeping antiquers awake at night is the tantalizing awareness that many old aircraft remain undiscovered. They sit forgotten in barns and sheds, awaiting discovery. This country is littered with undiscovered wreckages, some restorable, others useful for parts. The USAF agency keeping track of aircraft missing in the 48 contiguous states since 1941 has more than 3,000 unsolved cases on file.

The foreign scene is shrouded in mystery. Canada, Central and South America contain countless possibilities. The overseas theaters of World War II were long ago stripped of obvious tanks, guns, sunken ships, and aircraft by scrap dealers, yet they retain their secrets. One seemingly reliable account concerns a cave on a remote atoll. The Japanese aircraft in it were joined by U.S. types and the opening closed with explosives. Veterans recall seeing flyable airplanes towed into a shallow valley and bulldozed under on another island.

Most such rumors prove baseless—but not all. I'm looking at snapshots taken "somewhere in the Pacific." Among the hulks are A-20s, P-40s, B-25s, P-38s, F4Us, B-24s, two RAF Beaufighters and assorted transports—perhaps 40 airplanes in all. It appears that a number of flyable restorations plus others suitable for exhibit could be assembled.

Finding a relic is one thing; establishing ownership can be something else. Many governments, annoyed by unauthorized salvaging, have outlawed removing or even visiting wreckages. Alaska has declared all downed aircraft to be state property. One buyer of several surplus fighters in South America spent more getting his paid-for prizes released than it cost to ship them home. Getting signatures on documents at foreign ports can require considerable patience and payola. A domestic discovery might entail problems just as thorny. The lucky finder of an F4F in a Virginia swamp might learn he's recovered it *for* the Navy, which often maintains that anything it bought, no matter when, remains USN property.

One TIGHAR project is the investigation of reports that some German underground aircraft assembly lines were sealed shut (and doubtless mined) before invasion forces arrived. Can permission be obtained to open the eaves with demolition experts on hand?

Responsible restorers and historians continue the research, negotiation, and expense that lies behind the priceless relics we will one day see flying again or in museums. I admire them. They are perhaps as important in the scheme of things as those who built and flew the airplanes in the first place.

59

Silver in the sky

IN THE COOL OF EVENING we often sit outside enjoying the peace and quiet. Now and then a birdcall breaks the silence. Pelicans drift soundlessly across the tall pines in formations that would please the Blue Angels. Nothing could be more restful and serene.

Old sea captains built homes at ports so they could watch the ships. Airline retirees settle far from airports, at least large airports that handle the clamorous aircraft they once flew. We are 80 miles from the nearest jet terminal but not entirely removed from the scene. In the stillness at dusk we hear intermittent whispers of overhead traffic. It came as a surprise that turbine engines five miles up can be heard on the ground. But they can, and we spot silver sparkles way up where the sun still shines. I've counted half a dozen in 10 minutes.

When conditions are right for contrails the sky is laced with glistening tracks. It's languid down here, busy overhead. The thin white threads that trail behind tiny specks are Miami-O'Hare nonstops way up there. Junior crews in the lower levels leave cords and ropes of snowy vapor strung across the blue. Milk runs sometimes cross low enough for me to tell a Nine from a 737. I used to file 10,000 feet on short runs, as between Ft. Smith and Tulsa on clear spring evenings.

The aerial display stirs no envy. Flying for its own sake remains appealing and always will—take off, cruise, and arrival with riddles to solve en route. Even at this late date, I'd match skill and judgment against those lucky drivers up there. But the hassle that went with the job toward the last was frustrating, and it's worse today.

I follow a lofty speck moving fast, perhaps a 1011 or a Ten with three crew members up front and 14 in the cabin. Before they boarded, every one of them was screened for weapons by metal detectors and their bags were x-rayed. Sadly, it has become necessary to check passengers, but frisking pilots and flight attendants is preposterous.

Braniff 707-320C freighter en route to Saigon during Vietnam War.

At St. Louis I saw a security guard instruct a graying captain to open his suitcase and flight kit. She rummaged through them, found nothing suspicious, and motioned him on. I would like to have seen that tried with the crusty patriarchs I flew with 40 years ago. Screen us for weapons? You're not serious! Chief pilots would have agreed and that would have ended it. More than one inane proposal for increasing safety was quickly put to rest in those days. Times have changed.

"You'll have to admit that security checks have worked," says one captain with wry humor. "Since they were instituted, not one single airline pilot—not one—has hijacked his plane."

I ponder the cockpit atmosphere in one distant speck. Is it cordial or is a militant unionist paired with a strikebreaker? Are they giving full attention to flying or debating the specters of merger or bankruptcy? "I've had to tell my copilot and engineer knock off discussing B-scale pay," a captain told me recently. When pilots pay undue attention to differences during work they are not paying proper attention to the airplane. It's not widespread but any cockpit dissension is deplorable.

We did not escape distractions. During my last year our line was in desperate straits. The worst scenario appeared as a drastic retrenchment but not the total collapse that came. No "grandfather line" had gone under, no matter how serious its plight. We worried about matters that had nothing to do with the work at hand. That was not right. At least we endured none of the intercrew strife common in certain cockpits today. We were in it together to the bitter end.

The adversarial approach of the FAA in recent years has impeded safety throughout civil aviation. "I don't like to say it, but my priorities have changed," said one pilot. "When I'm on approach to LA, for example, my main concern is staying exactly on heading and altitude. If I stray, the controller will file a report. He's required to. I'd like to spend more time looking for traffic the controller may not see but I'm more worried about a violation."

The consequences of minor lapses are severe, whether or not they infringe on safety. One pilot cut a corner exiting a runway and knocked down a taxiway light. Quite properly he reported the goof. He was grounded for 30 days then required to requalify from scratch: ground school, simulator training, and rating ride. Reports are circulated about excessive penalties for petty mistakes. Clearly, a spotless record and good intentions count for nothing. Common sense is often replaced by harassment. The word is CYA.

A checkride is a stressful exercise, no matter how understanding the inspector. It can be made an ordeal that leaves the sharpest pilot looking foolish. There are stories of four-hour orals and rating rides. If an inspector cannot give a thumbs-up or -down after 90 minutes of 727 questions, he's browbeating. If he cannot complete a 727 rating ride in two hours, he needs more training himself.

The air carrier inspectors of my day were sharp, fair and reasonable. They were gentlemen. Make no mistake, there was no free lunch. You earned your ratings the hard way and passed your recurrent checks. A substandard performance required as many rerides as necessary, and if you couldn't cut the mustard you didn't get the ticket. It was understood that feds had a thankless job to do and we had strict standards to meet. A conscientious pilot had few problems with the FAA, or it with him, and he didn't dread disgrace because of a trivial mistake. The spirit of the law mattered, not the letter.

There was the rare exception—the petty tyrant who slipped into the system. This self-appointed crusader, whose piloting skills usually were less than outstanding, was out to shape up every pilot he examined. He was abusive. When pilots complained, managements interceded, and FAA officials listened. Misfits didn't last. But the sense of loyalty that once underlaid the pilot-management relationship on some lines is not what it was. A badgered pilot suffers in silence.

Happily, top FAA officials appear to recognize what pilots knew all along—the adversarial approach is unproductive—and want to restore the rapport of the past. I hope so. There's enough to think about in today's hectic environment without having to fear being bashed for a minor lapse that imperils no one.

The formula for maximum safety is simple: Populate the skies with airline crews sure of management support and sensible federal regulation, and assume that every other civil pilot has honorable intentions until he or she proves otherwise. The infinitesimal percentage that deliberately flaunts the law should be severely dealt with, including license revocation (not suspension), stiff fines, and prison time for flagrant violations. Let that news get around.

The crew aboard the airliner I'm watching might be required to provide urine specimens upon arrival. That's how random drug testing works; the order to make a "donation" is received upon arrival, without warning. Pilots argued that the program is unfair, unnecessary and unconstitutional—and got nowhere. Drug testing for flight crews has become part of airline life. Sobriety checks likely will follow.

The general aviation pilot who assumes the Fourth Amendment's ban of "unreasonable searches and seizures" guarantees his or her immunity from such random personal scrutiny should note that bus and truck drivers are not the only highway users being checked. The Supreme Court has ruled that police may set up sobriety checkpoints and require suspected private motorists to take breath tests. The possible extension to aviation is obvious.

It's dark. The silvery specks above have become winking lights that pass among the stars and fade from view. Having wrestled with some of the questions perplexing my former copilots and engineers and solved one or two, at least to my own satisfaction, I go in and turn on the television. It's almost time for "Wings" on cable. Tonight's subject is the DC-3. Should be good.

60

Milestones

WE WATCHED WITH MIXED FEELINGS as the Warrior taxied away. You're supposed to be as happy the day you sell as the day you bought, but we haven't found it that way with airplanes. We've owned four and missed every one from the moment it flew away.

The first was a slab-wing Fairchild 24K built in DC-3 times and every bit the same class act. Roll-down windows, big leather seats, chrome control stick, soft wide gear, it was a lovely souvenir of the golden age of private flying. Movie stars bought Fairchilds in the 1930s. Its six-cylinder, in-line Ranger came to life with the crackling snarl of a Spit's Merlin. It was a delight to fly and could be landed with little fuss in gusty crosswinds that kept modern trikes indoors. I'd buy it back at twice what we sold it for, but now 24s bring several times that. I hope old N20610 is still flying.

We upgraded to a G-35 Bonanza (my wife wanted something that "sits up straight" like a Buick), a sprightly performer that could almost keep pace with a DC-3. It was a fine cross-country mount, but you'd best pay attention. It picked up speed in a hurry when you stuck the nose over and needed TLC in chop. It could get an inept pilot into trouble before he knew it. We flew from Dallas to Kitty Hawk to see where it all began and to the far side of the moon down in the Big Bend country. (Go see for yourself if you think I'm kidding.) N4607D was comfortable, fun to fly and delivered surprising speed for its power. An American Airlines pilot bought it.

Having my professional wings clipped (for the sin of reaching 60) did not ease the itch, so we bought a Cessna 172. It was my second high-wing and the blind spot overhead was still worrisome. Visibility from airliner cockpits was never the best, but at least we didn't wonder who might be descending unseen above the wing. Otherwise the Skyhawk was admirable in all respects and good fun. An Englishman bought it and had it ferried to Biggin Hill, an old Battle of Britain fighter field near London. United States citizen N9621H became British subject G-BNXY.

Next we tried a Piper PA-28 Warrior. I never thought it particularly eye-pleasing but liked the way it handled. It had the solid feel that made an ex-jumbo driver feel right at home. It was roomy by lightplane standards and reasonably quiet. Landing was a piece of cake. The Warrior owner who can't pull off a smooth arrival nearly every time should not be allowed inside the airport fence. He needs more dual instruction. Or new glasses.

We went sightseeing along the Georgia Coast, watched shrimpers at work, and huge car carriers creeping up Brunswick Channel with 7,000 more Toyotas. We showed friends their homes from the air, steering clear of busy terminals, having no excuse to compound their congestion.

(The private airman who flies regularly, keeps up with procedures and has good reason to mix with airline, business, and military traffic at busy fields has every right to do so. Otherwise he is a nuisance and should stay away. His bumbling ilk frightened me badly more than once during my airline days.)

But I didn't fly regularly, didn't keep up, and had no pressing business at busy terminals so I stayed away. Flying without a purpose is a waste of fuel. Piper N38494 was—let's face it—an expensive toy. The hangar-insurance-maintenance clock ticked whether it sat or flew, so we let it go and watched it head for California. "It's another milestone," said my wife.

There are many milestones in the life of a career pilot, remembered moments meaningless to anyone beyond him and his family. Clear in recollection is a Canadian officer's quiet briefing that concluded, "Our country is at war. If any of you Americans have second thoughts, you are free to return home." No one spoke. "All right then, sign your enlistment papers—and thank you for volunteering."

Then the magic moment no pilot forgets—the first solo. I can hear the Kinner barking and feel the little Fleet rising from the grass as though it happened this morning instead of half a century ago. I remember looking back at the rear cockpit, unable to believe my instructor was on the ground watching. Weeks later I was "sent solo" in a howling 600-hp Harvard (known to cadets south of the border as the AT-6 or SNJ.) It was a handful for a teenager with 65 hours total time, but the ultimate toy.

Finally, "Wings Parade" in a chilly Ontario hangar. I stepped forward and snapped a proper palm-forward salute. Group Captain Nash pinned on the Royal Canadian Air Force "pilot's badge." That was indeed a milestone.

Some of the aircraft encountered thereafter are blurred in memory, others vivid. Each was a milestone. The C-47, C-46, and C-54 come to mind; in them I logged many hours. Transition to the C-54 involved hard book work and intensive flight training. Emerging from that schooling with a left seat ticket seemed more miracle than milestone. The first mad takeoff dash in a P-51 was indeed a milestone for one used to plodding transports.

Being hired by an airline after four discouraging years of trying was a milestone. Checking out as Convair captain 16 years and 12,500 hours later was another. Such events mean little to a pilot who hasn't been there, or a wife who hasn't agonized with a mate trying to absorb a semester of technical detail in three weeks. "How many ratings will we have after this?" asked mine while quizzing me on the 707's numbers.

She contributed more than she knew to my successful transition to that, my first jet, and later, monstrous "Fat Al." Her rides on my 747 trips to Hawaii, Guam, and Hong Kong were milestones we often recall. "Those were great times," she says today, and indeed they were.

On a fall day in 1964 we watched a Cessna 150 take off, circle, and land, our son Terry's first solo. Two years later he became the first pilot's son hired by our company. In 1978 he flew his first left-seat trip. His sister Kathy completed hostess training in 1972. Father-son crew pairings were discouraged but Kathy sometimes served in my cabin crew. The little dickens enjoyed startling my unknowing cockpit companions with, "Who's the dirty old man in the left seat?" After the inevitable lecture on protocol, we'd tell them.

Braniff never hired a better pilot or hostess then Len and Margaret Morgan's kids. Company history was made when Kathy married a pilot and Terry a hostess. Talk about milestones—never before had five in a family served as Braniff crew members. Between us we have seen much of the world during 70,000 hours of flight. So far, that is, for the two boys—well, they're boys to us—have years to go.

My final airline trip was a milestone both joyful and sad. I was as fit for duty the next day but no longer employable because of age. Mandatory retirement was easier to abide than the collapse of our great airline weeks later.

After collecting enough rejections to paper a T-hangar, having a piece accepted by *FLYING* was a remembered milestone. Few readers appreciate the sweat behind saleable writing. As one who has tried both flying and writing, I've found the latter is in some ways the more difficult. The urge to write, like the itch to fly, is hard to explain. I think it stems from reading. Anyone who reads widely on a familiar topic is apt to imagine he could say it better, or at least, differently.

An incurable bibliophile, I've assembled a fair collection of books, art, and memorabilia over the years. Collecting also has its milestones, those times when long sought-after titles were unearthed, or the control wheel of an old favorite was found in a boneyard, or a signed lithograph was spotted in a foreign catalogue. It took 30 years to put together a full run of *Jane's All the World's Aircraft*. Inside the first edition (1909) is a letter written by Fred Jane himself. I prize that.

The World War I shelves include a number of scarce items, some autographed and a few personally inscribed. The post-1918 books include references, histories, biographies, journals, pilot manuals, and logs. There's no hobby more intriguing than book collecting, but there is a time to collect and a time to dispose. A few treasured items I will keep but most of these things must go, and that will be still another milestone.

I believe the next will be Airplane No. 5—something simple without ILS, transponder, ADF, dual navs, reminders of when flying was work. Terry wanted me to buy his J-3 and I might have if it sat up straight like a Buick.

61

The image

THE MORNING MAIL BROUGHT A CATALOGUE of "pilot supplies." There are lots of useful things in it: charts, how-to videos, headsets, computers, logbooks. And neat things, such as sunglasses, flight clothing, plaques, chronographs. I'd almost forgotten chronographs. You weren't properly dressed without one in my salad days: six hands, rotating bezel, mysterious push buttons, black dial with luminous numbers you could read in the dark—sharp! It told the world you were an aviator.

Of course, you could fly anything built from Atlanta to Australia with the dashboard clock. You still can. The chronograph's prime purpose was to enhance the aviator image. Chronographs provoked snide comments by ground pounders about the "boys with the big wristwatches," which I will not repeat in full. Pure sour grapes. We Royal Canadian Air Force aircrew trainees had white flashes in our caps. The distinction drew sharp glances after it was rumored that the flashes identified servicemen with a communicable social disease. Again, pure sour grapes.

Here's a replica A-2 jacket with the "vintage effect of worn leather," which is sure to foster a 5,000-hour aura even if you soloed last Friday. And look at these Ray-Bans! This model with the green lenses and heavy browbar was essential to the Air Corps ensemble. Now and then you'd see some clown wearing the case on his belt. It was likewise bad form to wear a white scarf or earphones in Operations. Helmet and goggles were OK, but a scarf, earphones or Ray-Bans-on-belt marked the clueless.

The first thing to do with a new cap was discard the grommet, install a tarnished emblem from an old cap and bend the sides down. The resulting "50-mission crush" suggested experience, even hazardous duty. Now you can buy an identical cap already molded into shape. Add your own grease stains.

Clothes make the man unless all men wear the same clothes. We were as alike as links in a chain. The uniform was not enough. Personal demeanor, that established the image. For example, I overheard this from a domestic ferry pilot when asked, "How many enemy planes have you shot down, Lieutenant?" The shavetail closed his eyes and shuddered. "I'd like to forget all that," he said. It was personal demeanor that established the image.

Another newly-winged lad parried the slippery question with, "Must we talk shop?" Both had mastered the repartee that bridged the ticklish fact-fiction gap and preserved the "nothing can stop the Army Air corps" myth.

Humor was useful in sidestepping candor. Simply say, "How many airplanes have I brought down? Fourteen, and I was in all of them," then drift away during the laughter and find someone more interesting, like the unattached redhead across the room.

Liberties with unvarnished truth were not deemed dishonorable in distaff company. We weren't under oath, were we? One cadet carried it a bit far by explaining his Good Conduct ribbon was awarded for "extreme heroism in the face of incredible odds." However, his girl was humming, "He wears a pair of silver wings" before he finished primary.

The understatement for which the British are noted could amount to brazen overstatement. After the war, two RAF types raced to this side in a Meteor fighter, setting new records. When asked about navigation, one said, "Piece of cake, ecktually. We flew to Greenland and turned left." The more a feat was downplayed, the grander it seemed; however, any attempt to impress within RAF circles was unpardonable "line shooting." It "simply isn't done," a miscreant was warned.

The harmless tomfoolery of those long-ago times is amusing to recall. Flying began losing its mystique after the war. Pilots were no longer regarded as special people and, of course, they never had been. Soon almost everyone had a second cousin who flew for TWA. But was it ever fun while it lasted. Today's newcomer can't imagine what he missed.

Military training was tough and formation operations strictly disciplined, but a solo pilot flew pretty well as he pleased and few questions were asked if he brought his airplane back intact. When an "incident" required an explanation, a glib talker could often sell a less than precise account. Quizzing multiengine military (and airline) crewmen separately was the best shot at learning what really happened. Those days are gone.

First came the recording of air-ground radio chatter. It was for our protection, they explained; if there were questions, we'd know when who said what. The potential for self-incrimination was not at first apparent. I learned early on. We took off in a DC-7 and broke out on top to find the windshield filled by a head-on 707. The chance of collision was slight, but heated comment was exchanged—and taped.

At federal invitation, we listened to the playback. I heard myself blurting out things I could swear I never said. The captain and radar controller shook their heads at their own words. It transpired that we were cleared to make a left turn out. I'd copied it as a right turn and the busy controller missed the goof on readback. A lesson was learned by all and the file was closed; thereafter I handled my mike like a cocked Colt.

Knowing every word was recorded led to humorous comments. We were having our fillings shaken lose when the center requested a ride report. It bordered on severe, a description requiring a written report and airframe inspection; therefore my skipper responded, "I'd call it *extremely* moderate."

Radar made honest people out of cheats. The rude ace who routinely reported on short final when 10 miles out, scattering everyone in his path, had to clean up his act. Radar coverage was expanded until we were a blip on the screen literally from gate to gate. Deviations from flight plan required approval.

Next came black boxes tracing the airplane's speed, heading, altitude, and so on. These basics did not long satisfy the feds; later models noted 18 parameters, including the positions of control wheel, rudder pedals and flaps, g-loadings, roll and pitch attitudes, and much more. A crew's account was quickly collaborated—or refuted.

Then they installed cockpit voice recorder. This device self-erased everything more than 30 minutes old and could be wiped clean after landing, providing the airplane was on its wheels rather than its belly, and brakes were set. Manipulations were possible. Pulling the CVR circuit breaker froze an air-ground exchange a crew might wish to replay. (The feds soon made that illegal.) Prior to an unusual arrival—no flaps, and engine shut down, etc.—it was prudent to recite checklists and say nice things about the chief pilot during the last half hour.

The CVR was designed to preserve crew privacy, at least, we were told it was. The notion evaporated when we found phone jacks in the rear lounges of our Electras that were CVR-patched. We never learned who installed them or why. The spying device was thereafter rendered useless during preflight checks. I wonder if any airliners today include secret wiretaps.

In theory, the CVR was a good thing. When a crew died, the tape might help determine what went wrong. Many pilots felt investigators began paying more attention to CVR readouts than to probing wreckage and all of us were horrified to hear dead pilots' last words played on TV newscasts along with shots of crumpled dural. The media decided the public's right to know included its right to be ghoulishly titillated. Such coverage served no useful purpose, yet it continues, a shameful perversion of the CVR's intended purpose.

Self-made images are history. After tests and interviews, an airline employment office knows an applicant better than his or her mother. Thereafter a pilot performs beneath a microscope until retirement. His or her year includes two or three physicals, two checkrides preceded by oral and written exams, a company line check and unannounced federal checks. Every word spoken in flight is recorded, every mile flown watched. Not to forget random drug tests and frisking before every flight. No other profession is as closely monitored.

I thought fakery for fun and profit had vanished until I watched a boyish 727 second officer operate. A lady passenger called, "Oh, captain!" He doffed his cap, smiled broadly, and said, "At your service, Ma'am." He wore a chronograph to boot.

62

Kindred spirits

AFTER THE MEMORIAL SERVICE, Mick said, "I wish I could say I really knew him, but I cannot. He was both strange and brilliant. He was our brother, and a good brother, and we shall miss him." Strange is primarily defined as unusual, extraordinary. There was nothing usual or ordinary about our brother, David.

My earliest recollections of him are of a carrot-top who loved trains. The Central of Georgia Railroad linked our rural town with the outside world, in fact, it provided the only reasonable travel when rains turned red clay to slippery mud. There were no paved roads in Jasper County in those Depression days. Everything from coal to coffins came in by rail; cotton and peaches went to market on daily mixed trains.

I shared David's awe of steam locomotives and wanted to be an engineer, a notion quickly forgotten when a barnstorming OX-5 Travel Air landed in a pasture near town. Even after I said I didn't have three dollars and wouldn't have been allowed to ride anyway, its pilot let me look at the cockpit. My interest in trains waned. The glorious aroma of leather, dope, and gasoline was too much. David watched from a distance, unmoved.

We moved to Louisville and lived near the L&N's main line. David camped by the tracks, jotting down engine and Pullman numbers, and prowled the stations downtown. There was more for him to watch than I saw at Bowman Field where hours passed with little action. To show him the light, I bought us a hop in a slab-wing Stinson Reliant, unknown to our parents who would not have appreciated my missionary zeal. He agreed that flying was fun, then resumed his trackside research. "You've inhaled too much coal smoke," I said.

Boys who ached to sit in cabs or cockpits were expected to "grow up" and find respectable work. Railroading would do if it meant working at L&N headquarters—but flying? Oh, my dear!

David was 13 when I enlisted. In the half century that followed we rarely saw each other. We spent perhaps a month together all told, but we kept in touch and were closer than some brothers who live on the same street. In another sense I hardly knew him.

We considered ourselves hugely successful in being paid to do exactly what we wanted to do. Robert Louis Stevenson wrote in 1878, "I travel not to go anywhere, but to go. I travel for travel's sake. The great affair is to move." That's how we felt.

My interest was narrow; I wanted to fly airplanes, period. A larger picture intrigued him. He wanted to know everything about railroading: its history, its contribution to civilization world wide, its impact on modern life. He was hired at $37.50 a week by a small Milwaukee magazine, *Trains*, with a circulation of 39,000. Within five years he was editor and remained so for 34 years. I was hired by Braniff Airways at $47.50, with $10 deducted for a uniform. Neither of us doubted that grand years lay ahead, and neither would be disappointed.

We agreed that public transportation was by far the most fascinating of businesses but disagreed on its management. He deplored taxpayer funding of airports and airmail subsidies while railroads paid for their tracks and stations. I reminded him of the land grants that encouraged early rail expansion and that the global air transport network so vital in World War II was based on airline expertise. The fawning service accorded fares in DC-3 days amused him. "You won't be able to keep this up if and when you attract the mass market," he said, and we couldn't. After watching a P-80 fly in 1945 he wrote, "You'll be able to buy a jet ticket in seven years." I guessed 15 but he was right.

We watched and discussed dramatic changes—from steam to diesel power in his corner, from pistons to turbines in mine. The growth of air travel and shrinking of rail passenger service revolutionized our industries.

707-227 on final approach to Love Field, Dallas.

"David was the leading writer in the railroad field for more than three decades," said a friend. He was more, being an astute student of transportation in all its forms. His research took him and his wife around the globe and led to a hobby of collecting airlines, airliners, and places, all of them meticulously recorded in pocket notebooks.

From Nairobi and Nome and Sydney came picture postcards of foreign trains with teasing notes, "We rode an Alaska Coastal Grummman Goose today: Skagway-Juneau" or, "Add the Viscount to my roster: Zambia-Kenya on Central African." If a trunk or feeder flew it, David and Margaret probably rode it, from the Ford Tri-Motor ("Oh, but it's a fine thing to experience.") to the Concorde ("You've got to try this one, Len!"). He could get more on a postcard than most people can in a two-page letter.

Getting there the quickest way was secondary to adding another airliner or baggage tag to his collection; itineraries were planned accordingly. He and Margaret were aboard the ancient Sandringham flying boat Charlie Blair that flew to England in 1976. That publicized trip took 20 hours; they returned supersonically in three. Their logs eventually included almost 50 airliner types and airlines and hundreds of destinations.

After every trip came questions. If flaps decrease landing speed, why are they dropped for takeoff? Is the Concorde's cabin as wide as the Convair 340's? What's a step climb, a VOR, a PRT, a check valve? How many Electras were built? Was there a DC-5? I often had to dig for the answers yet I never once stumped him. His knowledge of railroading was encyclopedic.

David compiled his impressions in 1969 under the title, *FASTEN SEAT BELTS, The Confessions of A Reluctant Airline Passenger*, an insight into what passengers thought. What a shame that delightful little volume was not updated after his later travel adventures. He founded an excellent magazine for airline enthusiasts; it died after a year from lack of promotion.

We discussed writing. He had the uncanny knack of putting his thoughts on paper exactly right the first time. No revisions, no corrections. He made it look so easy. He was the pro, I the hack.

He loved ships. The passenger lists of the *Queen Mary*, *Queen Elizabeth*, and *France* on their sad farewell crossings included the Milwaukee couple. Typically, he wanted to be there when the curtain fell. I envied him those voyages.

Travel left him unjaded. He never lost his rapture for locomotives that snaked 120-car freights up mountain grades, great ships that rivaled great hotels, choosing dinner wine while moving at rifle bullet speed 11 miles up. It was not his nature to become blasé about such wonders.

Railroading was his life. He learned it hands-on, in steam and diesel cabs, cabooses, switch towers, presidents' private cars, at the same time maintaining a lively interest in flying and sea travel. A postcard from Long Beach of a restored steam locomotive read, "We stayed another day to see Howard Hughes' flying boat. Amazing!" He never missed an air show or TV documentary and sent a check to help finance Rutan's *Voyager*. Whenever his paper carried an aviation item, he clipped it for me, adding questions in the margin.

A stickler for detail and accuracy in what he wrote, his work reflected the drama and importance he saw in railroading. He appeared on network television, made talks, and was invited to the White House to watch President Johnson sign a railroad bill but he was a quiet, sensitive, private person who avoided attention. He spoke with his typewriter. Under his editorship *Trains'* circulation grew and today approaches 100,000.

Among other things, he kept the overall picture in perspective for me with his kidding. "Can't you guys stay away from those volcanoes?" when a 747 lost all power. "I thought 737s had brakes" when one overshot and stopped on railroad tracks. We in aviation have tunnel vision. Highly visible, we think we occupy the center of the travel universe whereas we own but a corner of it. David never let me forget it.

Illness forced his early retirement. He wanted to spend the winter here in our warmer climate. We would talk trains and airplanes and trips and faraway places and catch up on the 50 years gone by since we were kid brothers in a little Georgia town. At long last, we would get to know each other a little better. I think I looked forward to that more than he. But it was not to be. His condition worsened; a few days after Christmas he was gone and with him that wonderful chance. I never knew anyone else quite like him. I miss those cards and letters.

Index